WHAT IS AND WHAT OUGHT TO BE DONE

What Is
and
What Ought To Be Done
An Essay on Ethics and Epistemology

Morton White

New York   Oxford
OXFORD UNIVERSITY PRESS
1981

Library of Congress Cataloging in Publication Data

White, Morton Gabriel, 1917-
What is and what ought to be done.

Includes index.
1. Ethics—Addresses, essays, lectures.   2. Knowledge,
Theory of—Addresses, essays, lectures.   I. Title.
BJ1012.W49   170'.42   80-27852
ISBN 0-19-502916-X

Printing (last digit): 9 8 7 6 5 4 3 2 1

Printed in the United States of America

To
Herbert Wechsler

Preface

This study grew out of lectures delivered at the University of Oslo in the spring of 1978. Later that spring, I presented a paper on a related topic at the University of Zagreb, and in the spring of 1979, I presented a similar paper in Tokyo under the auspices of the Department of Philosophy at Tokyo University. I also defended one of the main theses of this essay in my Blanchard W. Means Memorial Lecture, given in the spring of 1980 at Trinity College in Hartford, Connecticut.

I warmly thank Dudley Shapere for discussing the central ideas of this study with me and for helping me to improve it after reading different drafts that were expertly typed by my secretary, Rose Murray. I also wish to express my deep appreciation to my wife, Lucia White, and to my son Nicholas White for making many helpful comments on these drafts. I am very grateful to Gilbert Harman for reading the penultimate draft and for his encouragement. It should not be inferred, however, that these generous readers bear responsibility for my views or that they agree with them.

Since I frequently comment on many numbered statements throughout this study, I have included a list of them for the reader's convenience; that list will be found at the end of the text.

Princeton, New Jersey   M.W.
August 1980

Contents

X   CONTENTS

WHAT IS AND WHAT OUGHT TO BE DONE

I

The Rejection
of Reductionism

1. *The Quest for Meaning*

G. E. Moore once wrote: "Ethical philosophers have . . . been largely concerned, not with laying down rules to the effect that certain ways of acting are generally or always right, and others generally or always wrong, nor yet with giving lists of things which are good and others which are evil, but with trying to answer more general and fundamental questions such as the following. What, after all, is it that we mean to say of an action when we say that it is right or ought to be done? And what is it that we mean to say of a state of things when we say that it is good or bad?"[1] Perhaps in recent times students of ethics have not been as concerned with the meaning of ethical statements as would appear from Moore's description, but it is fair to say that Moore's description truly characterizes the aim of many traditional moral philosophers as well as that of some contemporary philosophers. Because I do not share this view of the main task of ethics, I shall present an alternative view in this work. But before I do, I want to sketch the foundations of the view I do not accept.

Philosophers who espouse that view have concentrated on what is sometimes called a semantical question insofar as they try to discover what we mean when we make certain statements in what is called normative ethics. They hold that whereas ordinary men and professional moralists make substantive moral state-

1. G. E. Moore, *Ethics* (reprinted New York, 1949), pp. 7-8.

3

ments to the effect that certain actions ought to be done, the philosopher's task is to analyze the meaning of these statements made by ordinary men and professional moralists.

When philosophers regard their task as that of saying what we mean by ethical statements, they use the word "mean" in a technical way that deserves some comment. Such philosophers think that when they conclude their analytical labors they should present statements that are *synonymous with* the moral statements made by ordinary men. But what do such philosophers have in mind when they say this? In my opinion they have never been able to present an answer to this difficult question because their notion of synonymy is a very obscure one. And in order to see why, it is desirable to give an example of what such philosophers say they have in mind.

Since I am trying to convey some idea of a traditional philosophical opinion, I shall use a traditional illustration. Many philosophers have said that the statement "John is a man" is synonymous with "John is a rational animal", and they have regarded the connection between these two statements as different from the connection between "John is a man" and "John is a featherless biped". Why? Because, some philosophers say, the attribute expressed by "man" is identical with that expressed by "rational animal" whereas the attribute expressed by "man" is *not* identical with that expressed by "featherless biped".[2] It just happens to be a law of nature, they say, that all and only men are featherless bipeds and therefore that the terms "man" and "featherless biped" merely have the same *denotation* or merely refer to the same class of things. "All and only men are rational animals", they say, may be established merely by examining meanings or attributes whereas "All and only men are featherless bipeds" requires us to use our senses and to examine men. The same philosophers have emphasized that the alleged relation of synonymy between "man" and "rational animal" underlies the definition of the word "man"; they have also regarded the search for such a definition as the task of philosophical analysis.

Other philosophers of a related persuasion have not identified

2. See, for example, Rudolf Carnap, *Meaning and Necessity* (Chicago, 1947), esp. pp. 15-18. The reader who does not find these illustrations useful may, of course, substitute another pair. I understand that a featherless chicken has been bred, perhaps with dire consequences for this illustration.

analysis with a search for synonyms of verbal expressions but rather with one that leads to the conclusion that an extra-linguistic attribute or concept, such as being a man, is identical with the extra-linguistic attribute or concept of being a rational animal.[3] Yet one may question the clarity of both the notion of synonymy and the notion of identical extra-linguistic attributes[4] as used in much of the ethical thinking that has been carried out under the banner of analytic philosophy in the twentieth century. G. E. Moore was the most influential philosopher to espouse the analysis of attributes or concepts in the present century but he, ironically enough, came to the conclusion that the attribute of moral goodness is unanalyzable.[5] And, to compound the irony, Moore admitted toward the end of his life that he could not present an analysis of analysis itself and that he could not state with clarity what the task of analysis was.[6] Therefore, in my view, he also made the admission that he could not clearly defend his fundamental thesis that goodness is not analyzable.

Because the basic notions of the analytic or semantic approach to ethical statements and ethical terms are so wanting in clarity, I am convinced that this is not a fruitful approach to philosophical ethics and that it is better to adopt an epistemological approach by asking questions like "How do we know?" or "How do we justify our claims to knowledge?" Often epistemologists have confined themselves to raising this sort of question about knowledge gained in sciences like physics and mathematics, but it is obvious that we may raise the same questions about the knowledge we claim to have in normative ethics. Some philosophers who have adopted the analytic or the semantic approach in ethics have often done so out of a desire to answer an epistemological question, that is, to find out how we know that certain ethical statements are true. But they suppose that to discover how we know that a given action ought to be done we must give an analysis of what it means to say that an action ought to be done. In

3. See G. E. Moore, "A Reply to My Critics," in *The Philosophy of G. E. Moore*, ed. P. A. Schilpp (Evanston and Chicago, 1942), pp. 660-67. In the same volume, also see C. H. Langford, "The Notion of Analysis in Moore's Philosophy," pp. 321-42.
4. See my *Toward Reunion in Philosophy* (Cambridge, Mass., 1956), esp. Chapters 7-11 (reprinted New York, 1963).
5. G. E. Moore, *Principia Ethica* (Cambridge, England, 1903), esp. Chapter 1.
6. *The Philosophy of G. E. Moore*, p. 667.

my view, however, this is an unwarranted supposition. It seems to me possible to say something illuminating about the justification of ethical beliefs without venturing into the semantics of ethical language or the analysis of ethical concepts. In fact, I believe that it is necessary to bypass the murky notion of synonymy and the equally murky notions that we employ when we speak of the analysis of ethical attributes or concepts.

The obscurity of synonymy and of the notion of identical attributes not only deters me from trying to find synonyms for ethical terms or from asserting that ethical terms cannot be defined, but it also makes me skeptical of efforts to distinguish between so-called analytic and synthetic statements. Since that distinction is often based either on synonymy or on identity of attributes, if these notions are obscure, then the distinction between analytic and synthetic is at least as obscure. If we cannot be clear in asserting that "man" is synonymous with "rational animal" or that being a man is identical with being a rational animal, then we cannot be clear in asserting or denying that the statement "All and only men are rational animals" is analytic. In that case we cannot confidently hold that so-called essential predications like "All and only men are rational animals" are analytic whereas the truths of physics and other natural sciences are synthetic. It must also be borne in mind that the distinction between the analytic and the synthetic has been used to support an epistemological distinction between truths that are known a priori and those that are known a posteriori. If an analytic statement is said to be true by virtue of the meanings of its terms and if those meanings are attributes, then, some have argued, our knowledge of the truth of analytic statements is a priori. And if it be said in reply that the philosopher who distinguishes between analytic and synthetic statements need not assert the existence of meanings conceived as attributes but need only assert that certain verbal expressions are synonymous, that philosopher is in a difficult predicament so long as he is without a satisfactory account of synonymy. Whether such a philosopher rests on an obscure semantic notion like synonymy or on an obscure notion like the identity of meanings viewed as non-sensible entities such as attributes, that philosopher is in trouble.

Much of what I have said about this trouble is now widely

accepted by philosophers because of the salutary influence of W. V. Quine, and to these philosophers I offer my apologies for revisiting battlefields that may be all too familiar to them. However, it is not obvious that the lessons of earlier campaigns against analyticity[7] are now accepted by all moral philosophers. Thus Bernard Williams, the author of a relatively recent and well-received introduction to ethics, tells us that the term "good" in many of its occurrences functions as an attributive and not as a predicative adjective because, to use his language, a sentence such as "That is a yellow bird" *admits of the analysis* "That is a bird and it is yellow" whereas the sentence "He is a good cricketer" *cannot be analyzed* as "He is a cricketer and he is good".[8] Williams also says that whereas the sentence "This is a large mouse" *means something like* "This is a mouse larger than most mice", we cannot similarly analyze all expressions in which "good" appears attributively. It is difficult to see how the phrases "admits of the analysis", "cannot be analyzed", and "means something like" escape membership in the obscure circle that contains "synonymous" and "analytic" as used by earlier philosophers. When it is said that the sentence "That is a yellow bird" admits of the analysis "That is a bird and it is yellow", the implication is that the first sentence is synonymous with the second, just as Moore's statement "He is a brother" is supposedly synonymous with "He is a male sibling". The point is that almost seventy years after the appearance of Moore's *Principia Ethica,* Moore-like notions of synonymy and analyzability were still employed by a philosopher seeking to present a so-called logical feature of "good", namely, its being an attributive adjective and consequently very different from "yellow". In spite of appearances to the contrary,

7. W. V. Quine, "Two Dogmas of Empiricism," *Philosophical Review* 60 (1951), 20-43; reprinted in Quine's *From a Logical Point of View* (Cambridge, Mass., 1953), pp. 20-46 and elsewhere. Also see my paper, "The Analytic and the Synthetic: An Untenable Dualism," in *John Dewey: Philosopher of Science and Freedom,* ed. Sidney Hook (New York, 1950), pp. 316-30; reprinted in my *Pragmatism and the American Mind* (New York, 1973), pp. 121-37, and elsewhere. Nelson Goodman was also an ally in this campaign, as may be seen in his "On Likeness of Meaning," *Analysis* 10 (1949), 1-7, reprinted in Goodman's *Problems and Projects* (New York, 1972), pp. 221-30 and elsewhere.
8. Bernard Williams, *Morality: An Introduction to Ethics* (New York, 1972), pp. 41-42. The terms "attributive" and "predicative" are respectively applied to the uses of "good" and of "yellow" in the two illustrations.

then, contemporary analytic ethics continues to rely on some of the dubious concepts employed by earlier analytic moral philosophers.

Reliance on such concepts goes beyond the characterization of the adjective "good" as attributive. Williams maintains "that for many fillings of 'x' in 'that is a good x', an understanding of what an x is or does, and factual knowledge about this x—i.e., a combination of conceptual and factual information—is sufficient for one to determine, at least broadly, the truth or falsity of the judgment".[9] In other words, after arguing that the phrase "a good x" is attributive by using synonymy or allied notions, Williams goes on to argue for another thesis with the help of these same notions. This second thesis is best understood by considering what he says about his illustrative phrase, "a good can opener". He seems to hold that if we understand the term "can opener"—which for Williams amounts to understanding "what a can opener is"—we have what he calls "conceptual information". And such information is expressible in a statement that Williams calls a "conceptual truth". Notice, however, that a conceptual truth about "what a can opener is" is not very different from an analytic truth in the terminology of other philosophers. Moreover, if we have the conceptual information expressed in this supposedly conceptual truth, Williams says we are bound to understand the phrase "a good can opener". After asking: "Can we say that in phrases of the form 'a good x', the meaning of the whole is *essentially* determined by the meaning of what takes the place of 'x'?", Williams says that in the case of "a good can opener" the answer is affirmative. He writes: "If we consider functional descriptions of artifacts, such as 'clock' or 'can opener', . . . it does seem that if one understands these expressions (at least in the strong sense that one understands what a can opener is . . .), then one has understanding, within limits, of what a good thing of that sort is".[10] This is why Williams thinks that a combination of conceptual and factual information is sufficient for us to determine "at least broadly" the truth of the sentence, statement, or judgment "This is a good can opener". The conceptual information is presented in the statement of identity of concepts which tells us *what a can opener is;* the factual information is

9. *Ibid.*, p. 49.
10. *Ibid.*, p. 45.

that this shiny thing on the kitchen table illustrates the concept which is identified (in a conceptual truth) with the concept of being a can opener.

I think I have said enough to show that identity of concepts or attributes and synonymy—or some equally obscure cousins of theirs—have not left recent moral philosophy. Their continued presence is evident not only in the characterization of "good" as attributive but also in the view that we must know *what a can opener is* in order to understand the expression "can opener". According to this more recent view, knowing what a can opener is is not merely knowing that something is true of all and only can openers. We must know more than that. The parallel in the view of earlier analytic philosophers is that in knowing what a man is we go beyond knowing that all and only men are feather-less bipeds: we know that the attribute or concept of being a man is identical with the attribute or concept of being a rational animal.

My reference to the traditional effort to distinguish between the status of "Man is a rational animal" and "Man is a feather-less biped" gives me an opportunity to point out that Williams is refreshingly dubious about the attempt to "found morality on a conception of the *good man* elicited from considerations of the distinguishing marks of human nature". By contrast to what he says about the expression "a good can opener", he does not think that the meaning of "a good man" is "essentially determined" by the meaning of "man". In effect, he admits his inability to present what might be called the analysis of the attribute of being a man; and this inability is connected with a recognition that other characteristics, such as the capacity to make fire, to have sexual intercourse without regard to season, and to kill things for fun, are also coextensive with being a man. Williams shies away from the view that being a rational animal takes precedence over these other features of man in a way that would allow him to treat "a good man" as he treats "a good can opener". He sees that a good deal of evaluation goes into the Aristotelian selec-tion of rationality as *the* distinguishing mark of man.[11] But were he to accept something similar about *any* effort to present the analysis of a concept, he would shy away from saying that one of

11. *Ibid.*, pp. 63-64.

the many extensional equivalents of any general term may be singled out as that which presents "the meaning" of the term. Were Williams to agree with me about this, he would put "can opener" and "man" in the same boat and would avoid the use of questionable semantics in his treatment of *any* general term, including one that refers to artifacts like can openers and screw drivers.

Although I regard Williams's acceptance of the contrast between conceptual and factual truth as a throwback to some of Moore's views, I want to make clear that I am sympathetic with Williams's desire to avoid anti-naturalism in ethics by allowing some statements of the form "This is a good so-and-so" to be factual. However, my doubt about his distinction between conceptual and factual truth as well as my doubt about anti-naturalism grow out of a general suspicion of epistemological dualism. This suspicion not only leads me to feel uneasy about certain efforts to separate the method of testing so-called conceptual truths from that of testing claims of knowledge in the natural sciences but also to feel uneasy about similar efforts to separate the method of testing allegedly naturalistic descriptive statements from that of testing allegedly non-naturalistic normative statements about what ought to be done. Some philosophers say that normative statements are neither deducible from nor equivalent to those of natural science because normative statements ascribe properties which are said to be radically different from the properties ascribed in statements of natural science. Often the next step of such philosophers is to say that normative statements ascribe *non*-natural properties and that the possession of a non-natural property must be detected by a power which is sometimes called "intuition" and sharply distinguished from the intellectual powers employed by natural scientists. Some epistemologists of this persuasion hold that the intuition used in establishing fundamental moral truth is also used in mathematics, in which case they put mathematics and ethics on the same side of a barricade that divides both of them from natural science. But other epistemologists of a divisive turn of mind are not content with defending only one dualism since they hold that there is one method for establishing mathematico-logical truth, a second for establishing truth in the natural sciences, and a third to be used in normative ethics; they are, so to speak, trichotomists rather than dichotomists.

In the present work I do not intend to deal with all of the problems raised by the dichotomies or trichotomies I have just mentioned. In particular, I do not intend to deal in detail with the dualism between the analytic and the synthetic or with allied dualisms.[12] I shall concentrate instead on developing and defending a non-dualistic view of the relationship between normative and descriptive thinking.[13] In pursuing this goal I shall express my dissatisfaction with the distinction between the analytic and the synthetic where the expression of such dissatisfaction is relevant, but my main interest here is in expounding and defending a view which will avoid an epistemological dualism between the normative and the descriptive, and to do this without advocating what is sometimes called a naturalistic theory of the meaning of normative terms. My view does not require me to say what "ought" means or that it must be defined by using only terms of natural science. Being as skeptical as I am about efforts to analyze ethical concepts or to find synonyms for ethical terms, I shall avoid any effort to "reduce" normative ethics to natural science in a manner that relies on untenable contrasts between the analytic and the synthetic.

I want to say, however, that it is very difficult to show that the phrases "normative statement" and "descriptive statement" denote two mutually exclusive and exhaustive classes about which we can speak with confidence when we say that they are, or that they are not, tested in the same way. The statements "No man ought to lie", "Iago ought not to have lied", and "Smith had a right to defend himself" are typical normative statements from the point of view of those who think that they are established in a way that is fundamentally different from the way in which descriptive statements are established. By contrast, "Iago lied" and "Smith defended himself" are typical descriptive statements from this same point of view. But although this is the traditional contrast between the normative and the descriptive with which I *begin* my discussion, by the time I have finished it, I hope the

12. See footnote 4 above.
13. See *Toward Reunion in Philosophy*, pp. 254-58, 263, where I sketch the view that I shall here elaborate and try to improve.

reader will have been persuaded that when we think we are testing an isolated normative statement such as "No man ought to lie", we are putting on trial a conjunctive hybrid statement such as "No man ought to lie *and* Iago lied", a statement that seems to fall between the class of normative statements and the class of descriptive statements as usually specified.

Sometimes descriptive statements are characterized as statements that assert what *is* the case whereas normative statements are said to assert what *ought to be* the case or what ought to be done. But, from one point of view, thinking that an act or kind of act ought to be done is a species of thinking that something is the case, because it is possible to transform a statement like "He ought to have kept his promise" into "It is the case that he ought to have kept his promise". And, analogously, one may regard the denial of "He ought to have kept his promise" as "It is not the case that he ought to have kept his promise". Nevertheless, there is a difference between "He ought to have kept his promise" and "He kept his promise"; and this difference cannot be obliterated or disregarded by putting the expression "It is the case that" or "It is not the case that" before the statements just mentioned. The former of these statements says that a certain act *ought to have been* done by a certain person and the latter says that the same act *was* done by the same person, and that is why the former is here called a normative statement whereas the latter is called a descriptive statement even though each may be trivially equated with statements which assert that something is the case.

The usual distinction between normative and descriptive statements is most clearly seen when we contrast saying that *one and the same act* has been done and saying that it ought to have been done. It is obvious that we assert different things of the same act when we apply these different predicates to it. If we say "Brutus stabbed Caesar", what we assert is patently different from what we assert when we say "Brutus ought to have stabbed Caesar". For this reason, the contention that we cannot deduce "Brutus ought to have stabbed Caesar" from "Brutus stabbed Caesar", or the contention that they are not equivalent, is not very controversial. Most people and most philosophers are prepared to grant that from the statement that a given act has been done it does not follow that it ought to have been done. However, this is only one interpretation of the doctrine that "ought"-statements cannot be deduced from statements about what is the case, and philosophers have often wished to say much more when advocating that doctrine. They have not only maintained that you cannot deduce "Brutus ought to have stabbed Caesar" from "Brutus stabbed Caesar", the latter being a statement about what is the case and therefore descriptive; they have also gone on to argue that you cannot deduce "Brutus ought to have stabbed Caesar" from *any* statement about what is the case. For example, some have argued that you cannot deduce "Brutus ought to have stabbed Caesar" from the descriptive statement "Brutus's stabbing Caesar produced more pleasure than pain"; others have argued that you cannot deduce "Brutus ought to have stabbed Caesar" from the descriptive statement "God willed Brutus's stabbing of Caesar".

The stronger claim that no "ought"-statement is deducible from *any* "is"-statement or descriptive statement goes much further than the claim that a normative statement of the form "X ought to have done Y" cannot be deduced from the kind of descriptive statement that has the form "X did Y", where "X" and "Y" represent the same things in both statements. And it is the stronger or more general claim that is typically associated with what I have called a dualism between normative and descriptive thinking. For usually a philosopher who insists that a normative statement like "Brutus ought to have stabbed Caesar" cannot be deduced from *any* descriptive statement is prone to think of the whole class of normative statements as testable in a way which is fundamentally different from that in which descriptive statements are tested. Since he holds that you can never logically deduce any normative statement from a descriptive statement, he is likely to think that the method of establishing the truth of a normative statement must be radically different from the method of establishing the truth of a descriptive statement.

The idea that you cannot deduce a normative statement from any descriptive statement lies behind two of the more popular ethical theories of the twentieth century. One is the anti-naturalistic theory, already mentioned, which asserts that words like "obligatory" express *non*-natural attributes. It is obvious why this theory has found favor with some who deny that normative statements can be deduced from descriptive statements. If *no* descriptive statement like "Brutus's act produced more pleasure than pain" logically implies a statement like "Brutus ought to have stabbed Caesar", and if the predicate of the latter statement expresses *some* attribute, then it is alleged that that attribute must be very different from the "natural" attribute expressed by predicates like "produced more pleasure than pain", that is to say, predicates used in natural sciences like psychology. This alleged difference is then indicated by calling the ethical attribute "non-natural".[14] The other alternative to straightforward naturalism that has found favor in the twentieth century is the so-called emotive theory. According to one version of this theory, ethical predicates express *no* attributes at all and therefore have emotive rather than cognitive meaning. According to another version, ethical predicates have both cognitive meaning *and* emotive meaning.[15]

Since I discuss these views elsewhere,[16] I shall be brief in saying why I cannot accept them. I cannot accept anti-naturalism because the notion of a non-natural attribute has never been made clear by its inventors; and I cannot accept the emotive theory because it employs the obscure notion of cognitive meaning when it contrasts that with emotive meaning, which, in my opinion, is also obscure. But since it is sometimes thought that the only alternative to anti-naturalism and the emotive theory is ethical naturalism, I may well be asked whether I am an ethical naturalist. The answer is that I am not if ethical naturalism asserts that ethical predicates can be "analytically" reduced to certain predicates of natural science with which they are cognitively synonymous.

It is obvious, therefore, that in treating normative statements I am not a reductive naturalist who relies on the unclarified notion of cognitive meaning, an emotivist who relies on the unclarified notions of cognitive meaning and emotive meaning, or an anti-naturalist who relies on the unclarified notion of a non-natural attribute. Where, then, do I stand? The present study in

14. In *Principia Ethica,* Moore says goodness is non-natural whereas I have in mind the view that obligatoriness is.
15. Rudolf Carnap's *Philosophy and Logical Syntax* (London, 1935), pp. 22-26, contains a statement of the former version of the emotive theory whereas C. L. Stevenson's *Ethics and Language* (New Haven, 1944) contains the most thorough exposition and defense of the latter version.
16. See *Toward Reunion in Philosophy,* pp. 207-17; also Section 19 of the present work.

its entirety is my answer to this question, but it may be useful to present a glimpse of the direction in which (I hope) the reader and I may be moving together for some time.

My aim is to develop a considerably modified version of a doctrine advocated by Pierre Duhem in his reflections on the method of natural science, the idea that we do not test isolated individual statements but bodies, or conjunctions, of statements.[17] This doctrine—which I call "corporatism"—has also been advocated in a modified form by Quine, but his modifications and his uses of Duhem's doctrine are somewhat different from mine. Quine is mainly concerned to show that he can dispense with the need for an obscure distinction between the analytic and the synthetic[18] whereas I go beyond that by also trying to close the epistemic gap between the normative and the descriptive. I shall argue that we do not test normative statements individually or in isolation but rather as parts of bodies that include descriptive statements as well. And by moving in this direction I hope to illuminate the connections between normative and descriptive thinking while avoiding epistemological dualism and the pitfalls of some semantical or analytic ethical theories.

Before I conclude these introductory remarks I want to call attention to another concern of mine. I believe that epistemology contains normative statements and I shall try to use some of my conclusions about normative ethics in order to illuminate the nature of epistemology. We not only make moral normative statements in which we say, for example, that we ought not steal or that we ought to honor our parents; we also make epistemic normative statements in which we say, for example, that we ought not to believe that the earth is flat or that the moon is bigger than the sun. The statements "Einstein accepted the special theory of relativity" and "Einstein ought to have accepted the special theory of relativity" are respectively descriptive and normative statements, and therefore we should avoid a dualistic theory of their relationship just as we should avoid a dualistic theory of the relationship between "Einstein kept his promise" and "Einstein ought to have kept his promise". In my discussion

17. Pierre Duhem, *La Théorie physique: son objet, sa structure* (2nd edition, Paris, 1914), pp. 278-89.
18. See "Two Dogmas of Empiricism", *From a Logical Point of View*, pp. 42-46.

of epistemic normative statements, I shall be concerned in a general way with the relationship between descriptive statements made in the history or psychology of science and normative statements that are made in epistemology or in what is sometimes called the methodology or the philosophy of science.

Because the difference between saying what scientists *do* believe and saying what they *may* or *ought to* believe is analogous to the difference between the descriptive study of human behavior and the study in which the normative moralist engages, some philosophers hold that statements about what scientists do believe are established by using the naturalistic or empirical techniques of psychology whereas epistemic statements about what should be believed are established in a very different way. Other philosophers go to the other extreme and erase the distinction between the psychology and epistemology of science, but I try to avoid both of these extremes in my discussion of epistemic normative statements just as I try to avoid analogous extremes in discussing the epistemology of moral belief. I have placed the latter discussion before the former in the present study because philosophers who have engaged in discussions of ethics for thousands of years have provided some guidance for the treatment of normative epistemology. Therefore, once I have presented my views on moral normative statements, it will be easier to make my views on epistemic normative statements more comprehensible and more convincing, or so I hope.

II

Corporatism, Science, and Ethics

2. *Epistemological Corporatism: Total and Limited*

Having presented some idea of the epistemological dualism I reject in dealing with ethical beliefs or statements, and having characterized the sort of reductionism that I do not wish to employ in avoiding that dualism, I now want to lay the groundwork for my own view of how to establish normative beliefs. It is modeled on what I have already labeled "corporatism", but my corporatism has certain historical antecedents in the epistemology of purely descriptive belief.

I have already indicated that I intend to develop a modified version of an idea presented by Pierre Duhem in his reflections on the method of natural science, the idea that we do not test statements or beliefs individually but rather bodies of statements. And I have also indicated that Quine has been influenced by Duhem to adopt a related view. Neither, however, has dealt with the relationship between normative and descriptive beliefs, my main concern in the present study. Therefore, in this section I shall present the views of Duhem and Quine, as well as my own, about the testing of bodies of purely descriptive statements or beliefs so as to prepare the way for my treatment of bodies of beliefs some of which are descriptive and others normative.

I shall begin by saying something about Quine's views about the testing of descriptive beliefs, views with which I am in partial agreement. Quine maintains in his paper "Two Dogmas of Empiricism" that we do not test our beliefs individually but rather that we test the total body of our beliefs. According to him, our total body of beliefs is a tool that we use in the organization and prediction of sensory experience, and in using it we—as a rule tacitly—use all branches of it: logic, mathematics, ontology, natural science, social science, geography, history, and even descriptive common sense as illustrated in statements like "There are brick houses on Elm Street". The fact that we use all branches of science in this way means that any statement in any branch may be changed or rejected in the face of what Quine calls a recalcitrant experience that might be thought to count only against some individual statement that is ostensibly being tested by itself. It should be borne in mind, therefore, that there are for Quine three major elements in the situation: (i) the scientific thinker, (ii) the total body of science which he uses as a tool for the purpose of working, to use Quine's phrase, a manageable structure into the flux of sensory experience, and (iii) the flux of sensory experience itself.[1]

Part of the philosophical thrust of Quine's view may be seen by considering an example in which a pharmacist counts the pills in one bottle containing only red pills and does the same for a bottle containing only blue pills.[2] He concludes that there are 434 bottled red pills and that there are 292 bottled blue pills. Accepting the statement that $434 + 292 = 726$, he expects and predicts that if he were to spill the contents of both bottles of pills on the table and were to count them up while they were on the table, he would find 726 in all. But suppose that after spilling them, he counts and finds only 725. Experience has become recalcitrant, Quine would say; the unexpected has occurred and the pharmacist must patch up his thinking. Quine also holds that in the course of this patching, the pharmacist may change his logic and arithmetic in an effort to accommodate the recalcitrant experience. By contrast, other philosophers would claim that since the pharmacist may test the arithmetical equation "$434 + 292 = 726$" by the "non-empirical" methods of the positivist, the Platonist, or the Kantian, the pharmacist may not

---

1. *From a Logical Point of View,* pp. 42-46.
2. I choose this rather simple illustration in the hope that it will serve better than a complicated scientific illustration in helping me to clarify certain epistemological doctrines.

patch up his thinking by amending or rejecting this arithmetical equation. He is obliged, some philosophers would say, to find his error in the realm of so-called empirical statements. For example, he may show that a pill rolled off the table, or that he counted up the red pills or the blue pills wrongly after spilling them out of their respective bottles, and so on. But he is not empowered to touch logic and mathematics.

Even though Duhem was a forerunner of what Quine calls holism, Duhem would have required our pharmacist to find his error in a limited domain of belief. For Duhem did not extend what Quine calls holism to the totality of our beliefs. Duhem did not maintain that the logic and pure mathematics a physicist uses are subject to revision by an appeal to experience, and therefore Duhem would not have held that the pharmacist in our example may amend or abandon the arithmetical statement that $434 + 292 = 726$ while searching for a way in which to deal with his recalcitrant experience.[3] Duhem thought that mathematical statements should be established in a way that is different from that in which we establish physical statements.[4] Therefore, if we call Duhem a holist, we must bear in mind that his wholes are smaller than Quine's. For Duhem the whole of *physics* might hang in the balance in a so-called crucial experiment,[5] but not the whole composed of all our beliefs, as Quine would have it. Under the circumstances, I think it would be better to use the word "corporatism" to designate the doctrine that unites both Duhem and Quine, just because the word "holist" if applied to Duhem might suggest that he held that any one of our beliefs is subject to change or denial when a physicist encounters a recalcitrant experience. The point is that the word "holism" suggests that the biggest whole of all, the whole of belief itself, is tested when we might think we were testing only one belief, whereas the word "corporatism" allows for the possibility that bodies of belief which fall short of the whole of belief are tested.

3. For example, Duhem writes: "la pure logique n'est point la seule règle de nos jugements; certaines opinions, qui ne tombent point sous le coup du principe de contradiction, sont, toutefois, parfaitement déraisonnables," *La Théorie physique*, p. 330. Duhem believed that only some of our judgments may be rejected merely because they are self-contradictory.
4. *Ibid.*, pp. 285-89.
5. *Ibid.*, pp. 278-85.

If we call both Quine and Duhem epistemological corporatists, we may then distinguish between them by calling Quine a *total* or holistic corporatist and Duhem a *limited* corporatist.

On one issue I agree with Duhem rather than with Quine although I agree with both of them that often when a person appears to be testing an isolated belief, he may surrender other beliefs in the face of a sensory experience that might appear to disconfirm only the supposedly isolated belief. As I understand Duhem, we distinguish beliefs which we might reject instead of the belief that is ostensibly under fire from beliefs which we would not regard as relevant while trying to accommodate a so-called recalcitrant experience. If I were to arrive at the conclusion that my car is now in my garage and then found upon opening the garage door that it did not appear to be there, I might, as Quine suggests, plead hallucination and hence give up my assumption that I was in a normal state when I looked into the garage. But, like Duhem, I think that some beliefs played no part in supporting my conclusion that my car was in the garage and therefore were not candidates for rejection or revision when my car did not seem to be in the garage. For this reason I should call myself a *limited* corporatist, agreeing with Duhem that we test *limited* bodies of belief.

However, when I examine the limited bodies of belief we do test, I do not conclude that they all come from one discipline, as Duhem seems to have thought when he examined physics. Here I agree with Quine because I think that the beliefs which make up the bodies we test are heterogeneous from a disciplinary point of view and that any one of this heterogeneous collection of beliefs—mathematical, logical, ontological, or physical—may be amended or surrendered in the face of recalcitrant experience. That is why I hold that our pharmacist might have amended or surrendered a belief of arithmetic that he used while dealing with his pills. In sum, I believe that often we really test *limited bodies* of belief when we appear to be testing only one belief while I also agree with Quine that these bodies of belief are heterogeneous and that any member of such a body might be altered or abandoned in order to deal with an experimental setback. It should be realized, however, that the limitation and the heterogeneity in question will vary from body to body of belief. For example, in some cases, one might consider abandoning the

Pythagorean theorem because it *does* play a part in supporting some other belief ostensibly under fire, in which case the Pythagorean theorem would be *within* the body of belief tested. And in some cases a whole class of beliefs—for example, those of ichthyology—might not be in the heterogeneous body that would be altered or abandoned in the event of a scientific crisis in astronomy.

Corporatism as I understand it must be distinguished from another view that might be confused with it. Philosophers of all persuasions would grant that a pharmacist might count the pills on his table, conclude that there are 725 there rather than the expected 726, and then decide to check his addition of 434 and 292, or ask a colleague whether there were indeed 434 pills in one bottle and 292 in another, or look under the table for a pill, or call up his wife to ask whether she had the impression that he was losing his capacity to count, and so on. All philosophers recognize that those who balance checkbooks go through things like this periodically. But such attempts to discover where we made a mistake or mistakes are different from what the corporatist has in mind. The typical bank-depositor's search for a mistake would not lead him to reject a previously accepted truth of arithmetic like "$3 + 2 = 5$", but corporatism implies that he might reject even that. Certain philosophical empiricists, however, would not permit him to do so because this would violate their belief that a sensory *experience* cannot logically dictate the acceptance or the rejection of an arithmetical or a logical statement.

Corporatism may best be understood if formulated as the view that sentences like the following one are elliptical: "The pharmacist put on trial his belief that if he were to spill all of the pills from his two bottles onto the table, there would be 726 pills on the table". And why is such a sentence elliptical? Because it does not make explicit all of the beliefs of the pharmacist that would be placed on trial once he began—as some might think—to test the mentioned belief by itself. For example, it would not make explicit that the pharmacist believes that there were 434 pills in one bottle and that there were 292 in the other. Nor would it make explicit that he believes that $434 + 292 = 726$. Moreover, it would not make explicit certain laws of logic that the pharmacist believes, laws that he might have assumed in any empirical testing or trial of his belief about what would happen if he were to spill all of the pills from his two bottles onto the table. As he tried to get from his single belief to the experiences that would confirm or disconfirm that belief, he could be asked: "Are you not assuming that there were 434 pills in one bottle?", "Are you not assuming that there were 292 pills in the other bottle?", "Are you not assuming that $434 + 292 = 726$?", and so on. And as he answered "Yes" to these and similar questions, he would be showing that the original sentence about what he was putting on trial was elliptical, showing that more was on trial than the statement about how many pills would be on the table were he to empty both bottles.

Therefore, my argument for corporatism when only descriptive beliefs are involved is as follows. Generally speaking, when one seems to be testing an isolated belief, other beliefs play a part in one's effort to check the ostensibly isolated belief by appealing to sensory experience. And these other beliefs therefore stand in jeopardy just because a believer or tester has a right to reject or alter at least one of *them* rather than the belief ostensibly on trial if the believer or tester should not have certain predicted sensory experiences. Therefore, these other beliefs are also on trial. One may elaborate on this argument and one may raise questions that I shall try to answer later about the nature of the thesis of corporatism but at this stage I do not think it necessary to say much more in defense of corporatism. A philosopher who denies corporatism would have to hold that one is always bound to focus on the belief ostensibly under scrutiny and to reject or amend *it* in the event of an experiential set-back. But I cannot see how anyone could hold this position without supposing that all of the other beliefs that figured in his testing were neither revisable nor rejectable on the basis of sensory experience. The idea that *some* of the other beliefs were neither revisable nor rejectable on empirical grounds might be accepted by a corporatist like Duhem, perhaps on the untenable assumption that logical and mathematical statements are a priori or analytic, or on the equally untenable assumption that they must be established by examining super-sensible entities such as universals are thought to be.

In rejecting a dualistic corporatism such as Duhem's I find myself in agreement with Quine, as I have said. But since I have also disagreed with Quine's view that *every* one of our beliefs is on trial in *any* experiment or test, I want to say just a little more about that view of his. Quine may be led to hold that view by imagining that all of knowledge has been logically systematized. Therefore, it may be that when he considers a limited piece of reasoning by a scientist who is testing a body of statements containing a logical law which may be regarded as a premise in the argument, he immediately thinks of the logical law as part of a deductive system of logic. And so he might think that one who considers surrendering the logical law should realize that we *might* make a change somewhere else in the deductive system of logic rather than surrender the particular law that governs our particular argument. Generalizing this, Quine may hold that whenever a limited argument of the kind I focus on is considered, we must imagine that each non-logical premise also brings into jeopardy every statement in the deductively systematized discipline from which the premise comes, as well as every statement from any other discipline that is presupposed by that discipline. And perhaps Quine thinks that this process will ultimately bring *every* belief into jeopardy. By contrast, I am inclined to say that at any given moment some beliefs are not components of deductively systematized disciplines and that challenging such beliefs does not imply that all beliefs in the disciplines containing them may be challenged instead and therefore put in jeopardy. That is one reason why I do not think that whenever we seem to be testing an isolated belief we are testing the *totality* of our beliefs. This does not mean, of course, that the conjunctions we test need contain very few conjuncts. Such conjunctions may be of any size whatever, and if it could be shown that *every* belief is put in jeopardy by a *given* faulty prediction, I would accept Quine's total corporatism. However, so long as it has not been demonstrated that all of our beliefs are interconnected parts of one system, some of those beliefs need not be touched by the shock-waves of a faulty prediction.

3. *Some Concessions Made and Some Objections Answered*

I wish to make a minor concession to those who think that some statements are tested in isolation and that we do not always test *bodies* of belief. The concession is that the tested conjunction itself might be regarded as a single statement which is tested in isolation. After all, it logically implies a conclusion the rejection of which leads by logic to the rejection of the conjunction itself. For consider the pharmacist and his pills once again. He may be represented as arguing in the following way: "If $434 + 292 = 726$; *and* there were 434 pills in one bottle; *and* there were 292 pills in a second bottle; *and* all of the pills in both bottles were spilled on the table, then there are 726 pills now on the table. But it is not the case that there are 726 pills now on the table. Therefore, the conjunction: '$434 + 292 = 726$; *and* there were 434 pills in one bottle: *and* there were 292 pills in a second bottle; *and* all of the pills in both bottles were spilled on the table' is not true". Clearly the pharmacist is rejecting a single isolated statement when he rejects the quoted conjunction before he begins to alter it so as to bring his views in accord with experience. But this is a conjunction of some of the statements which comprise the tested body and that is why my limited corporatism might have been expressed by saying that when we think we are testing a single statement we are usually testing a conjunction composed of that statement and other statements. We test more than the statement that is most conspicuously on trial in an experiment. We test a conjunction of *it* and other statements. And the fact that statements other than *it* may be surrendered in the face of an experiential set-back shows that a conjunction is on trial. It also shows that this conjunction may be altered by amending or abandoning statements other than the conjunct that seems at first to be on trial all by itself.

Now, however, I must face another objection, one that is best stated and answered by using a different illustration. Suppose that a conjunction of scientific statements leads to a conclusion such as "This piece of paper is red". Imagine that the piece of paper is litmus and that we have dipped it into an acid. In that case we may be putting on trial, according to the corporatist, a conjunction of statements that will include "This piece of paper is litmus", "Whenever a piece of litmus paper is dipped into acid, the paper will turn red", and logical statements. But the question concerns the testing of the conclusion, which, if true, will confirm the conjunction of premises and which, if false, will disconfirm it. Is this conclusion about the redness of the paper tested all by itself? My answer is that it is not. When testing *it*, once again we test a conjunction such as the following: "This piece of paper is red; *and* I am a person of normal vision; *and* I am looking at this piece of paper in white light; *and* whenever a person of normal vision looks at a red thing in white light, that thing will look red to that person". And my main point here is that even a conclusion about the objective color of a physical object—a conclusion which is supposed to confirm or disconfirm a body of descriptive statements—will itself be put on trial with other statements and will therefore not be tested in isolation. To get from such a conclusion of a hypothetico-deductive argument to sensory experiences while testing the premises of such an argument one must make assumptions that are put in jeopardy along with the conclusion itself.

Finally, let us bear in mind that each conjunction that is put on trial will contain logical laws when our reasoning is made fully explicit, and therefore they are also conjuncts in the tested body of statements. They too may be amended or surrendered in the light of our desire to accept a conjunction that is in accord with experience. Of course, viewing logical laws in this way is opposed by philosophers who hold that logical statements are impervious to the vicissitudes of experience and who therefore subscribe to a dualism between the methods of testing logical truth and empirical truth. But such philosophers might be somewhat mollified if they recognized that my corporatism does not lead to the view that logical laws by themselves logically imply something about experience. If our pharmacist's argument contained as premises the logical laws he uses in his argument, those laws would not by themselves logically imply any statement about how many pills were on the table or about what seemed to be on the table. The *conjunction* of those logical laws and his other statements would logically imply such statements but that conjunction would not be a logical truth because at least one of its conjuncts would not be a logical truth. For example, the conjunctive statement "Either a statement is the case or it is not the case; and there are 434 pills in the bottle" is not a logical truth even though its first conjunct is.

In recapitulating what I have said so far about limited corporatism as applied to descriptive statements, I shall represent the doctrine in a schematic way. Since corporatism asserts that we usually test a conjunction of statements containing a given statement S when we might think that we are testing S itself, let us represent such a conjunction as follows:

(a) S and T and U and V and W and X and Y.

When testing such a conjunction as (a), we may derive a logical consequence, say Z, in an effort to test (a). But Z itself, I have said, may also be tested as one member of a conjunction. Let us represent that conjunction as follows:

(b) Z and A and B and C and D.

Now let us assume that (b) logically implies statement E, which reports an experience. Let us also assume that we do not have the experience which it reports. For example, E may report a certain experience of redness but in fact it seems to us exactly as if we were seeing something green. In that case, we have had a recalcitrant experience and E is false.

Now if E is false, then (b) is false. If (b) is false, then at least one of its conjuncts is false and should be denied. Let us suppose that we deny only Z. In that case we must deny (a), which implies Z. And when we deny (a), we must reject at least one of its conjuncts. Let us suppose that we reject component T rather than component S, the statement we might have regarded in a non-corporatistic frame of mind as the only one we were permitted to reject. By rejecting T we exercise the Duhemian epistemic right that corporatism gives us.

So far I have laid most emphasis on this right to reject a conjunct like T instead of one like S when experience disappoints our expectations, but there will, of course, be many occasions when experience is in accord with our expectations and therefore *confirms* a conjunction like (a). In many instances of such confirmation, however, one statement like S may loom in the mind of the tester as *the* statement that has been confirmed even though he knows that in the event of an experiential set-back he might have rejected T rather than S. Thus, although the fact that a piece of paper looks red may confirm the *conjunction* "This piece of paper is red; *and* I am a person of normal vision; *and* I am looking at this piece of paper in white light; *and* whenever a person of normal vision looks at a red thing in white light, that thing will look red to that person", the person who is doing the testing may have a special interest in testing the statement "This piece of paper is red". In that case all of the other conjuncts will serve only as links, so to speak, in a logical chain that will help him to decide whether to accept "This piece of paper is red". Being in such a situation is quite compatible with my version of corporatism. The scientific tester who focuses his attention on one statement may well conclude that "This piece of paper is red" is what he has confirmed *on this occasion* simply because he accepts all of the other conjuncts. This is not to say that they have been previously established in isolation. On the contrary, they too will have been established as conjuncts of other conjunctions from which they will have been extracted and used as reliable assumptions in the deductive argument that helps the tester confirm "This piece of paper is red". Like the principle of conservation of energy or some principles of logic, they will have earned their credentials in some earlier corporatistic test. That is why they will have been relegated to the background as stable or reliable assumptions; and, for this reason, the experience that confirms a conjunction of these stable assumptions *and* the hypothesis in which the scientist is particularly interested at a given moment may sometimes be thought of as confirming the latter alone. The stable assumptions are taken for granted and therefore not acknowledged as beneficiaries of the confirmation gained by the hypothesis that happens to be in the forefront of the scientist's mind.

Since I have spoken, as Quine does, of descriptive science as a tool which organizes our sensory experiences, I had better say something about this phrase. It is obvious that if descriptive science organizes a person's sensory experiences, then there are occasions on which these experiences are not organized. So let me begin by trying to describe a situation in which my sensory experiences are not organized, in which they lack a manageable structure. Right now, as I write these words, I seem to hear a noise which I can only describe by saying that it sounds like a noise the reader might produce for himself by pronouncing "sh", "sh"—as if trying to hush someone into silence. The noise seems to be coming from my right, and so I look to my right and see that a window is open. It is about 4:45 P.M. on a warm day late in November and darkness has almost fallen, but there is enough light for me to think, after the "shushing" stops, that I see a dark patch moving through the air exposed by the open window. The dark patch moves to the left and, as it does, I hear a noise which is different from the "shushing" noise. This second noise will be harder to describe by likening it to the pronunciation of certain words. It seems to be a grating noise. Soon after moving to the left, the dark patch moves to the right and I continue to think I hear a grating noise. But as soon as the patch disappears, I seem to hear the old "shushing" noise again. Throughout all of this interval I am having sensory experiences of sight and sound which, I should say, are not organized. I cannot link them with each other. I do not know what is happening outside; I do not have a hypothesis that will organize my experiences.

Suppose, however, that I develop a set of hypotheses which I think will organize them. I come to believe that there is a squirrel outside my window and that it is crawling on a screen when I hear what seems to be a grating noise. As the squirrel moves to the left, I see a patch and also hear the grating noise. When the squirrel reverses direction and moves to the right, I continue to hear a grating noise while the squirrel is *on the screen,* but when, I hypothesize, the squirrel leaves my field of vision it moves onto an ivy vine on the invisible brick wall to the right of the window, at which time the "shushing" noise begins. The squirrel is rustling the ivy leaves. Once I adopt this conjunction of beliefs about a squirrel that moves back and forth from screen to ivied wall, I have organized my experiences. I have made it possible to link some of them with others and therefore to predict and explain phenomena that were previously unpredictable and inexplicable.

In this situation my experiences have been organized or linked by a *conjunction* of beliefs or statements. If called upon to articulate the organizing conjunction, I would state the premises that lead logically to "Now I seem to hear a 'shushing' noise", "Now I seem to see a dark patch", "Now I seem to hear a grating noise", and so on. Among these premises there would be generalizations connecting the squirrel's motions with the noises on the metal screen and also with the noises on the leaves; there would also be logical laws. Having this set of premises in hand, I may say that I can now link sights and sounds that I could not previously link. Of course, if I were to advance some other set of premises that implied statements about experiences contrary to experiences I had, I would be in a different position. My body of beliefs would fail to link my experiences, and I should in that case scrutinize my premises in order to link the experiences I did have.

The foregoing is intended primarily as an account of how limited corporatism may be applied to descriptive statements alone, an account which is preliminary to my treatment of the epistemology of normative belief. Therefore, I have presented a discussion of the relationship between an acceptable body of descriptive belief and experience, a discussion which will be sufficient for my purposes in what follows. A fuller discussion would require fuller treatment of those features of an acceptable body of belief which go beyond organizing or linking experiences in the manner described. Thus, when two competing bodies of belief both link the same experiences, other considerations must be adduced as a basis for preferring one of the bodies. For example, simplicity is such a consideration, and so is the fact that the preferred body does not dislodge what William James called the "older truths" from their august status in our thinking. I shall have more to say about this subject when I come to discuss the statements of epistemology, but before I do that I want to turn to my principal concern in this work, which is to apply limited corporatism to bodies of belief that contain normative ethical beliefs as well as descriptive ones.

4. Corporatism and Ethics

Neither Duhem nor Quine discusses the testing of normative ethical beliefs when expounding their versions of corporatism, but I believe that we can better understand the relationship between normative and descriptive thinking by recognizing that nonconjunctive normative ethical beliefs may sometimes be components of limited, heterogeneous bodies of belief that are tested in a corporatistic manner. In normative ethics we test a limited but diversified body or conjunction of beliefs and sometimes such a body of beliefs may lead to a moral conclusion the rejection of which may prompt us to alter or reject a previously held nonmoral belief. So, whereas Quine divides the scientific situation into three major elements: (i) the descriptive scientific thinker, (ii) the body of science that the thinker uses as a tool for organizing or rationally linking sensory experiences with each other and (iii) those sensory experiences themselves, I, when I deal with the epistemology of normative moral belief, use the triad: (i') the normative moralist, (ii') the body of descriptive and normative moral beliefs that he uses as a tool for rationally linking certain sensory experiences and certain emotions or feelings, and (iii') those sensory experiences and emotions.

Now for some clarification by way of illustration. Suppose that a mother has taken the life of a fetus that she has been carrying, and suppose that we present the following argument:

(1) Whoever takes the life of a human being does something that ought not to be done.

(2) The mother took the life of a fetus in her womb.

(3) Every living fetus in the womb of a human being is a human being.

Therefore,

(4) The mother took the life of a human being.

Therefore,

(5) The mother did something that ought not to be done.

Once we grant that after denying (5) we may deny the conjunction that implies it, we should recognize that the denial of this conjunction is equivalent to the alternation of the denials of its conjuncts. Consequently, we *may* amend or surrender a law of logic like that which gets us from (2) and (3) to (4); an ethical law like (1); or a descriptive statement like (2), (3), or (4). And such amendment or surrendering constitutes an alteration of our orginal body of beliefs.

Since we need not deny or alter (1), our normative principle, we are free to deny or alter some other kind of statement, one that is descriptive. If we deny (3), thereby undermining the argument to (4) and (5), we exchange our old conjunction of beliefs for a new one by denying a descriptive belief—and (3) is descriptive because it contains no "ought" or "may"—but it should be emphasized that we deny that descriptive belief because we reject (5), a normative conclusion that follows from our former assumptions. Thus the denial of (5), which denial is also a norma-

tive belief, may play a part in determining what descriptive beliefs appear in our little system, since the denial of a descriptive
belief (3) is also descriptive. In short, we have altered our system
by adopting the *descriptive* view that not every living fetus in
the womb of a human being is a human being because we have
adopted the *normative* view that it is not true that in killing the
fetus the mother did something that ought not to be done. And
this is analogous in a certain respect to a physicist's amending or
rejecting a previously accepted logical belief because of certain
data of quantum mechanics. The right to alter one's logic in response to certain experiences in physical experiments is analogous
to the right to alter one's description of an act in response to
one's moral assessment of the act.

I wish to protect what I have said from certain misunder-
standings. First of all, I am not saying that the denial of (5) logically implies the denial of (3). That would be equivalent to saying that (3) itself logically implies (5), which is absurd. From
"Every living fetus in the womb of a human being is a human
being" you cannot deduce "The mother did something that ought
not to be done". Analogously, as we have seen, if our pharmacist
were to reject a logical statement or an arithmetical equation because he rejected the statement that there were 726 pills on his
table, that would not justify saying that the logical statement or
the arithmetical equation itself logically implied that there were
726 pills on the table. What logically implies the conclusion in
both cases is a conjunction of premises and not any one of them
taken by itself. Therefore, since some philosophers, for example,
Hume, think it is fallacious to *deduce* an "ought"-statement from
an "is"-statement,[6] I want to emphasize that I have not licensed
this. I have merely said that if a conjunction containing descriptive and moral statements—like the one in the reasoning from (1)
through (5) above—logically implies a moral conclusion which is
denied, we may alter the conjunction by surrendering or amending one or more of the premises, and that the surrendered or
amended premises may be either descriptive or moral.

Moreover, when we consider a conjunction of statements
some of which are normative and others descriptive—say, the conjunction of (1) and (4)—it seems hard to say that the conjunction

6. *A Treatise of Human Nature,* ed. L. A. Selby-Bigge (Oxford, 1888), pp.
469-70.

is normative or descriptive. It certainly seems difficult to say that it is descriptive if it contains a normative conjunct. And it seems equally difficult to say that it is normative if it contains a descriptive conjunct. For let us bear in mind that a normative statement is, as I have identified it, one which asserts what ought to be done, what ought not to be done, or what may be done. If, therefore, the conjunction of (1) and (4) asserts what ought not to be done *and* what has been done, we may well wonder whether the conjunction is normative. To be sure, it *implies* a statement which says what ought not to be done, but then it also implies a statement which says what has been done. The conjunction, then, seems to fall into a third category; and, for this reason, the corporatistic idea that we test only such conjunctions seems not only to show that we cannot make an exclusive and exhaustive division of descriptive and normative statements but also that such a division is pointless once we regard such a heterogeneous conjunction as the linguistic expression or belief we test. The division appears applicable only to non-conjunctive statements like (1) and (4). It seems to break down when we try to apply it to the conjunction of (1) and (4) because such a mixed conjunction says what ought to be and what is.

However, instead of recognizing a class of statements that are neither normative nor descriptive, we may say that a conjunction can be a normative proposition in spite of containing a descriptive conjunct or in spite of implying a descriptive conjunct. For example, if we say "He did it and he ought to have done it", we may argue that our conjunction is normative. And if we say "Brutus's act of stabbing Caesar ought to have been performed", it might be held that our statement is normative in spite of implying the descriptive statement that Brutus stabbed Caesar. If one adopts this approach, one may say that a mixed conjunction of propositions, some of which are "ought"-propositions and some of which are "is"-propositions, is an "ought"-proposition. In that case, the conjunction of premises in the argument from (1) through (5) would be regarded as an "ought"-proposition and I would formulate the view I am defending by saying that if we deny the *normative* conclusion (5), we may deny the *normative* conjunction of premises leading to (5). However, I would add that if we deny the normative conjunction of premises to accommodate our denial of the normative conclusion (5), we may deny a descriptive component of that conjunction, namely, (3). In that case we have denied a descriptive statement in the wake of denying a normative one.

Another way of looking at the matter is to say that normative ethics presupposes some part of descriptive science if only because individual acts must be described in order to be judged, and because a moral principle asserts that any act which is described as being of a certain kind ought, ought not, or may be done. The relationship between normative science, then, and descriptive science is not unlike that between physics and logic or arithmetic. We use arithmetical expressions in order to express physical truths and we make arithmetical assumptions when deducing some physical truths from others. Similarly, we need psychological and sociological concepts and truths in order to express normative truths and to derive some normative truths from other normative truths. We must describe what we are prescribing or proscribing. The reasoning from (1) through (5) illustrates the latter point. The phrase "takes the life of a human being" is a descriptive one which is needed for the statement of normative principle (1); and the descriptive statements (2) and (3) are assumed in order to permit the deduction of normative statement (5). Therefore, we cannot test a normative belief like (1) without putting in jeopardy certain descriptive beliefs when we engage in normative thinking of the kind illustrated by the argument from (1) through (5).

In general, then, the moralist must keep in mind all of his premises and make alterations while viewing every such premise as a component of a conjunction that he alters with due attention to *features of the whole conjunction*—for example, its simplicity—that are beyond a statement-by-statement examination of it which is conducted under the illusion that every premise can be checked by itself without attention to the heterogeneous body of which it is a part. Therefore, we may reject the view that we examine one realm of being to check our logical principles, a second to check our so-called factual statements in a practical argument, and a third to check our moral principles. That is to say, we may reject the widely held view that we should inspect universals in order to establish a law of logic; that we should conduct a very different investigation of physical or psychological entities in order to show that a human being has performed a certain action; and that we should conduct yet a third kind of investigation by examining so-called natural and non-natural universals in order to show that a moral principle is true. Philosophers who hold this view would have us look at classes— which are universals—with the mind's eye in order to establish the logical statement that if one class is included in another and an individual is a member of the first class, then the individual is a member of the second class. They would have us use the body's eye in testing the statement about concrete individuals that one human being has killed another. And they would have us study the connection between the natural relation of killing and the non-natural attribute of being a forbidden act to see whether a moral law against killing is true. In this way such philosophical advisers might hope to divide the process of assessing a moral argument and thereby avoid corporatism. But such efforts are vain. The epistemological thesis that there are different ways of knowing truth cannot rest on a trichotomous ontology which is itself exceedingly vulnerable. Nor can it be supported by appealing to a semantic trichotomy favored by those who distinguish between the analytic sentences of logic, the synthetic sentences of natural science, and the emotive sentences of ethics. Such a trichotomy rests, as I have already said, on a dubious theory of cognitive and emotive meaning. However, I would not wish to argue that merely because this view is defective, limited corporatism is acceptable. If the distinction between the analytic and the synthetic is not acceptable as a basis for distinguishing between the method of testing mathematico-logical truth and that of testing truth in natural science, and if certain ways of distinguishing the method of testing truth in natural science and that of testing truth in ethics are also not acceptable, then corporatism looms as an attractive alternative to the trichotomy I have mentioned. But there is more than that to be said in its favor, as I have tried to show.

So far I have said that descriptive science should link or organize sensory experiences and that normative ethical science should link or organize sensory experiences and what may be called moral emotions or feelings. I have also illustrated this view and argued for it. It is worth adding that I do not mean that in assessing pieces of reasoning in these sciences we think of nothing but their capacity to effect such linkage. Other considerations will enter into our assessment, for example, the duty to disturb our system as little as possible and the duty to construct a simple system. As a consequence, there are moments in our thinking when some of these other epistemic duties, so to speak, conflict with our primary epistemic duty as thinkers—that of linking experiences with each other or experiences with moral feelings. For example, we may occasionally refuse to believe our senses, as the saying goes, when they engender conflict with the dictates of conservatism or simplicity. Thus, the physical principle of conservation of energy may be legitimately saved by appealing to the epistemic principle of conservatism when some sensory experiences go against the principle of conservation of energy. And the scientific preference for smooth curves may be allowed to prevail over what we observe in a laboratory by appealing to the epistemic principle of simplicity.[7] But the normative rules of scientific thinking are often ranked, just as the normative rules of ethical behavior are; and it would seem hard to deny that the main rule of scientific thinking is that scientists ought to link sensory experiences with each other and that the main rule of ethical thinking is the obligation to link sensory experiences with each other and with moral feelings.

7. See Nelson Goodman, *Ways of Worldmaking* (Indianapolis, 1978), *passim*.

III

Linking Sensory Experiences and Moral Feelings

5. *Normative Conclusions and Moral Feelings*

Now that I have presented the central thesis of my corporatistic view of the relationship between normative and descriptive beliefs, I want to deal with possible objections to that view and with certain alternative views. This will help clarify my position and give it further support.

Since we sometimes reject normative conclusions of certain arguments by appealing to our moral feelings, I want to consider some ways of doing so. I begin with an illustration which may be more useful than the one that runs from statement (1) to statement (5) above. Consider the following statements:

(6) Every act which is a lie is an act that ought not to be performed.

(7) The prisoner's act of saying yesterday at 4 P.M. "My regiment went north" is a lie.

From them we may derive the following counterpart of (5) in the earlier illustration:

(8) The prisoner's act of saying yesterday at 4 P.M. "My regiment went north" is an act that ought not to have been performed.

In my view, such an argument might be offered by a moral judge of some logical cultivation who has certain experiences and moral feelings. The premises of the argument are presented in an effort to justify the moral judge's acceptance of statement (8) and in this way to link those experiences and feelings. And the situation of this moral judge is analogous to that in which a purely descriptive scientist might find himself when he explains why a piece of litmus paper is red by citing the law that acids turn litmus paper red and the singular statement that the piece of litmus paper in question has just been dipped into an acid. I say "analogous", but no more than that. For it must be remembered that the descriptive scientist tries only to link sensory experiences with others whereas the moral judge tries in addition to link sensory experiences with moral feelings. The descriptive scientist's argument contains no normative statements. The descriptive scientist, by using the linguistic structure he has built, will be able to connect some of his sensory experiences with others, and in this sense organize them; the moral judge, by using the linguistic structure *he* has built, will be able to connect his sensory experiences with his feeling of obligation. Using a mechanical metaphor rather than the biological metaphor of organization, one might say that the descriptive scientist builds conceptual bridges that allow him to move from sensory experiences to sensory experiences whereas the moral judge builds conceptual bridges that not only get him from some sensory experiences to other sensory experiences but also from sensory experiences to his moral feelings. I should add, however, that the moral judge does not causally explain his or anyone else's moral feelings. Doing so would be the task of a descriptive psychologist. When a descriptive psychologist offers a causal explanation of some person's feelings, he makes no normative statements and therefore does not try to link his experiences with any of his own moral feelings. The descriptive psychologist links sensory experiences with others in trying to give a purely causal explanation of some other person's feelings. By contrast, the moral judge links sensory experiences *and* moral feelings when he presents the kind of argument illustrated by (6), (7), and (8).

I have formulated the above illustrative argument so as to bring out the fact that moral principles state that certain kinds of *acts* ought, ought not, or may be performed. I also wish to emphasize that (7) is a descriptive statement by the standards of philosophers who contrast non-conjunctive descriptive and normative statements. It is just as descriptive as the statement that a living fetus in the womb of a human being is a human being. If a lie is, as the *OED* says, "a false statement made with intent to deceive", then (7) is not a *moral* statement.[1] I grant that sometimes moralists use the word "murder" so that it, as some philosophers might say, contains the notion of being morally wrong as part of its *meaning*. For such moralists "The prisoner committed murder" is a moral statement. But the analogous view of "The prisoner told a lie" is not forced upon us. Statement (7) need not be interpreted as saying that the prisoner *wrongfully* made a false statement. It merely asserts that he made a false statement with the intent to deceive, and neither "intent" nor "deceive" is here used as a normative word. They are used as psychological words in describing an act which is prohibited in statement (8).

Before dealing in detail with the question whether we may reject (8) by consulting our moral feelings about an act, I want to prepare the way for my answer by considering a parallel mode of rejection in purely descriptive science. Let us recall that a purely descriptive body of belief is supposed to predict sensory experience and that it may do so by implying a belief which is confirmed by sensory experiences. Therefore, a body of descriptive beliefs functioning as a set of premises may be accepted because a conclusion that is deduced from the set of premises receives confirmation in sensory experience. The confirmation that the conclusion receives from experience will be credited to the account of the body or conjunction of premises; and the disconfirmation that the conclusion receives from experience will be debited to that same account.

Bearing this in mind, let us suppose that a body of descriptive scientific premises leads to the conclusion that a certain powder will be white. However, the powder does not seem to be white. Following a pattern discussed earlier, we test our conclusion as part of the conjunction: "This powder is white; *and* I have nor-

1. See Immanuel Kant's definition in his essay, "Über ein vermeintes Recht aus Menschenliebe zu lügen," *Kants Werke* (reprinted Berlin, 1968), Band VIII, p. 426. One translation renders the passage I have in mind as follows: "Thus the definition of a lie as merely an intentional untruthful declaration to another person does not require the additional condition that it must harm another," Kant, *Critique of Practical Reason, and Other Writings in Moral Philosophy*, translated and edited by L. W. Beck (Chicago, 1949), p. 347.

mal vision; *and* I am looking at the powder in white light; *and* whenever a person of normal vision looks at a white thing in white light, that thing will appear white to that person". This conjunction is put to the test when it is supplemented by the logical law that helps the whole conjunction imply that the powder *appears* white to me at a given time. But, as I have said, the powder does *not* appear white at that time, and so I have a recalcitrant experience that requires me to decide which premise in the body of premises implying "This powder is white" should be changed or rejected. Suppose that I immediately deny that the powder is really white because I do not reject any other statement in the conjunction just mentioned, such as "I am a person of normal vision", or "I am looking at the powder in white light", or "Whenever a person of normal vision looks at a white thing in white light, that thing will appear white to that person". Once I reject "This powder is white", I am required to deny or amend at least one statement in the body of descriptive scientific premises which implies that the powder *is* white.

Turning now to our example of ethical reasoning, we might argue that we may treat statement (8) above in a similar way. In that case we would assume that there is some recalcitrant moral feeling or emotion that is analogous to the recalcitrant experience in the example about the powder, a feeling or emotion that would justify the rejection of statement (8), "The prisoner's act of saying yesterday at 4 P.M. 'My regiment went north' is an act that ought not to have been performed". Is there such a feeling, one that stands to statement (8) as the appearance of the powder stands to the statement that it is white? If there are such feelings or emotions, then we may argue that sometimes when we have one of them, we may reject the normative conclusion (8) and then search among the premises that lead to (8) in order to see which should be rejected or altered. Furthermore, that search may lead us to decide to reject the descriptive statement (7), that is to say, to deny that the prisoner told a lie. Alternatively, we may deny (6).

We must, however, in modeling our rejection of (8) on our rejection of "This powder is white", avoid the idea that a moral feeling of *any* person in *any* situation could justifiably contribute to the overturning of (8). And this means that (8) itself would be a conjunct in a tested conjunction such as the following: "The prisoner's act of saying yesterday at 4 P.M. 'My regiment went north' is an act that ought not to have been performed; *and* I am a normal person; *and* I am contemplating the prisoner's utterance under circumstances that correspond to viewing the powder under white light; *and* when, under such circumstances, a normal person contemplates an utterance which ought not to be made, that person will feel obliged not to make the utterance".

The view I have just outlined requires us to fill in the blank, as it were, which refers to a normal person and to normal circumstances that correspond to a normal viewer's looking at a physical object in white light—and different cultures and different persons may fill the blank differently. The view also requires the existence of a feeling which stands to the predicate "ought not to be uttered" very much as a sensory experience of whiteness stands to the predicate "is white" in the examples I have just used. In other words, the view requires us to believe that there is a feeling or emotion the having of which will confirm *the conjunction* that contains (8) just as there is a sensory experience the having of which will confirm the corresponding *conjunction* that contains "This powder is white". Moreover, if we do not have this feeling on the appropriate occasion, we shall have to reject the conjunction that contains (8); and if we are moved to retain every other conjunct, we shall have to reject (8).

I believe in the existence of moral feelings such as this with as much confidence as I believe that there is a sensory experience to which we appeal when we attribute a color to a physical object. I cannot see how we can deny the existence of such a feeling and, more particularly, how we can deny that when we sincerely *say* that we feel obliged to do something of a certain kind we in fact have the sort of feeling in question. And if I were asked to give an argument for believing that there are feelings of obligation, I should respond as I would if asked to give an argument for the view that we have sensory experiences of red or green.

At first it might seem that feeling obliged to do (or to refrain from doing) something is in this respect fundamentally different from having an experience of red. Why? Perhaps because the word "red" seems less complicated than the corresponding ethical predicate, "act that ought not to have been done". Consequently, some philosophers think that although it is sometimes true to say "It seems to me exactly as if I were seeing red" and thereby to report a sensory experience, it is never true to say "I feel exactly as if I were obligated to keep that promise". Yet I do not think that a statement beginning with "It seems to me exactly as if I were seeing . . ." can only be true when the three dots are replaced by a relatively uncomplicated expression like "red". Indeed, we can sometimes replace the three dots by as complex an expression as "a Cadillac" and still form a true statement. For this reason I think that we do report feelings or emotions of obligation when we truly say: "I feel exactly as if I were obligated to keep that promise" just as we report "guilt-feelings" when we truly say "I feel guilty about not having kept my promise". The complexity of these statements is no obstacle to their expressing feelings, just as the complexity of "It seems exactly to me as if I were seeing a Cadillac" is no obstacle to its expressing an experience. On the basis of such considerations I am prepared to say that there are feelings of obligation to do certain things just as there are sensory experiences of red, and that both may be cited in testing conjunctions of the kind under discussion.

The idea that there might be a recalcitrant feeling about an action (regarded in a certain way) which would prompt us to reject a conclusion like (8) and then to reject a premise that might be a moral principle or a descriptive statement runs counter to a view which has been advanced by W. D. Ross. He contrasts the method of ethics and the method of natural science by saying that "the moral convictions of thoughtful and well-educated people are the data of ethics just as sense-perceptions are the data of a natural science".[2] In addition, Ross holds that it is "self-evident that a promise, simply as such, is something that *prima facie* ought to be kept".[3] We can see therefore that Ross would regard his moral principle about keeping promises as incapable of appearing in a hypothetico-deductive argument that might lead to the rejection of the principle. For when Ross says that it is self-evident that we have a *prima facie* duty to keep a promise, he means to say of this principle and of others like it "that they come to be self-evident to us just as mathematical axioms do".[4] And this in turn means that once they are seen to be self-evident, they cannot be overturned in the way that Ross thinks the hy-

2. W. D. Ross, *The Right and the Good* (Oxford, 1930), p. 41.
3. *Ibid.*, p. 40.
4. *Ibid.*, p. 32.

potheses of natural science can be overturned by certain sense-perceptions. Therefore, my view is very different from that of Ross since I think that we test a conjunction of a logical law, a moral principle, and a descriptive statement that is not logical when we view a practical argument as a hypothetico-deductive system. Such a conjunction will be tested in part by reference to its capacity to link "perceptions" and moral feelings; and moral feelings are not moral *convictions* or beliefs.[5] Therefore, feeling as well as perception may help to overturn or revise a mixed body of beliefs.[6]

I want to stress that my corporatism does not imply that the predicate "ought not to be performed" *means the same as* "would be felt by a normal person under normal circumstances to be one that ought not to be performed". Nor, of course, does it imply that "objectively white" means the same as "would *appear* to be white to a person of normal vision when viewed in white light". It is important to stress this in order to show that my corporatism does not involve reductionism and therefore does not rely on the notion of synonymy. If I assert that some powder is objectively white *and* that I am a person of normal vision *and* that I am looking at the powder in white light *and* that whenever a person of normal vision looks at an objectively white thing in white light, that thing will appear white to that person, then—if I also assume a certain amount of logic—I can infer that the powder will appear white to me without committing myself to saying that a certain statement is synonymous with "The powder is objectively white". And the same goes for an analogous statement about the prisoner's act. I am not giving an "analysis" of what it means to say that it ought not to have been performed. In each case I am giving an account of how it may be put on trial along with other statements which are on trial with it.

However, if one were to adopt the naturalistic view that "ought not to be performed" *does* mean the same as "would be felt by a normal person under normal circumstances to be an act

5. Thus I reject two views of Ross: that moral beliefs are the data of ethics whereas perceptions are the data of science, and that moral principles are self-evident.
6. Insofar as I appeal to feelings, the view I defend may be compared with that of Bertrand Russell in "The Elements of Ethics," *Philosophical Essays* (London, 1910), pp. 1-58; reprinted in W. Sellars and J. Hospers, *Readings in Ethical Theory* (New York, 1952), 1-32. See Part III of Russell's essay.

that ought not to be performed", then each statement in a practical argument would be equivalent to a naturalistic descriptive statement. If one adopted this view as well as corporatism, then denying or revising premise (7) because one denies (8) would raise none of the objections that are levelled by some philosophers against denying or revising a descriptive statement because one denies a normative conclusion that is *not* said to be synonymous with a naturalistic statement. According to such a naturalistic approach, our earlier practical argument would be equivalent to the following:

   (6') Every act which is a lie is an act that a normal person under normal circumstances would feel obligated not to perform.

   (7') The prisoner's act of saying yesterday at 4 P.M. "My regiment went north" is a lie.

Therefore,

   (8') The prisoner's act of saying yesterday at 4 P.M. "My regiment went north" is an act that a normal person under normal circumstances would feel obligated not to perform.

Every statement in this argument is descriptive. Therefore, denying (7') after denying (8') would be completely parallel to denying that one had dipped a piece of litmus paper into acid after denying that it had turned red while one resolutely maintained that litmus paper always turns red when dipped into acid. I say this only to show that a reductionistic naturalist who accepts corporatism can *easily* hold what I have been advocating whereas, by rejecting reductionistic naturalism, I have a harder time in doing so. However, it is evident that in rejecting reductionistic naturalism I do not deny that there are important similarities between a moralist's thought-processes and those of a natural scientist.

   In calling attention to the similarities between what natural scientists do and what moralists do, I realize that some philosophers might say that physicists *explain* their data whereas moralists *justify* theirs. Such philosophers might say that a natural scientist who points out that acid turns litmus paper red and that a piece of litmus paper has been dipped into acid *explains* the sensory experience of red whereas a moralist who points out that no one ought to lie and that the prisoner lied does not explain moral feelings. Rather, such philosophers might say, the moralist *justifies* or *offers a defense* of a feeling of obligation. But this linguistic observation does not count against my view: I am quite prepared to acknowledge its correctness while maintaining what I have maintained. The moralist's argument will do something for his feeling of obligation *and* his experiences that is analogous to what the natural scientist's explanatory argument does for his experiences alone—namely, organize them or link them. Perhaps some philosophers say that the moral argument *justifies* feelings whereas the scientific argument *explains* sensory experiences because they think that we can alter feelings of obligation at will but cannot alter experiences of red at will. In that case, such philosophers might also say, we give *reasons* for having certain feelings but do not give reasons for having certain sensory experiences. Even if this distinction between what we do in the case of feelings and what we do in the case of sensory experiences were valid, I do not think it would militate against my view about what moral arguments and scientific explanations have in common. If one has certain experiences and certain feelings about the prisoner's act, then the linguistic structure represented by the reasoning from (6) and (7) to (8) will link these experiences and feelings in the same way as one's experiences about the litmus paper and the acid will be linked with one's experiences of red by the chemist's linguistic structure. This is compatible with calling the former structure a moral justification and the latter a causal explanation.

6. *Feeling an Obligation and Feeling That One Is Obligated*

Since I have said that there are feelings of obligation that are analogous in an important respect to sensory experiences, I want to deal with a question that requires me to make a distinction of some importance. It might be pointed out that we often make statements like the following: "I feel that the prisoner's mendacious act of saying yesterday at 4 P.M. 'My regiment went north' is an act that ought not to have been performed". And it might be stressed that the opening words of this statement are "I feel *that*", with strong emphasis on the "that". Furthermore, it might be said that the expression "I feel that" could be replaced in this context by the expression "I believe that". If so, a philosopher might argue, the speaker is not reporting full-blooded feelings or emotions of the sort that we have when we are in a fit of rage or in a state of pain. He is reporting a *belief* that the prisoner did something that he ought not to have done.

I think that Hume rightly held that belief is a sentiment but I shall not press that point here. Although I am prepared to speak of *feeling* that a proposition is true, I have something different in mind when I speak of a feeling of obligation. It is not a feeling *that* someone has an obligation; rather, it is a feeling *of* obligation. And just because this feeling of obligation is not expressed by such cognitively oriented words as "I feel *that*", it resembles a feeling of pain or rage in a certain respect and *is* analogous to the sensory experience of seeing red. A feeling of obligation may not be as vivid or as strong as a feeling of (severe) pain or of (intense) rage but, on the other hand, it can be very strong indeed in persons with a highly developed sense of duty.

A feeling of duty or obligation is not unlike a feeling of sadness in being non-cognitive. And a feeling of sadness is not reported by saying "I feel (believe) *that* I am sad"; it is more directly reported by saying "I feel sad", just as a feeling of obligation is reported by saying "I feel obligated", or "I feel dutybound". Similarly, a sensory experience of seeing red is not reported by saying "I see that an object before my eyes is red". As soon as one uses the phrase "I see that", one intellectualizes the situation in a way that I wish to avoid in the case of sensory experience and moral feeling. And so I avoid (whenever I express myself properly) saying "I feel that I am obligated" when I want to refer to the feeling of obligation that is analogous to sensory experiences.

The distinction I have in mind may be made clearer by recalling the distinction between what William James calls "knowledge of acquaintance" and "knowledge-about",[7] as well as the distinction that Bertrand Russell makes between "knowledge of

7. William James, *The Principles of Psychology* (New York, 1890), Volume I, p. 221.

things" and "knowledge of truths".[8] James says "I know the color blue when I see it, and the flavor of a pear when I taste it", and adds that he cannot impart his acquaintance with this color and this taste to anyone who has not already made the acquaintance himself. Russell says that we are acquainted with sense-data and adds that we can know something *about* a sense-datum such as a shade of color—for example, that it is rather dark—which is different from the acquaintance we have with the shade of color. James also says that "we *have*" that with which we are only acquainted. Now the feeling of obligation (*about* which I have been speaking) is something that I *have* on occasion, just as I have the experience of blue and the feeling of sadness on occasion. And this feeling of obligation which I have, and with which I am acquainted, is not to be confused with any belief. It is not to be confused with the belief that a certain act is obligatory even though such a belief is often *called* a feeling. And it is certainly not to be confused with the belief that I have *it*. In other words, if anyone should take it into his head to say "I feel that I have a feeling of obligation", it should be clear that although his word "feel" may be equivalent to "believe", his word "feeling" is not equivalent to "believing".

In distinguishing between feeling obligated to do something and feeling *that* someone is obligated to do something, I have more than an idle interest in clarifying what I advocate. If it is thought that the feelings to which I refer are moral *beliefs,* it might be mistakenly thought that I subscribe in some measure to the coherence theory of truth. For it might seem that in testing a set of premises in a moral argument, I never look beyond a moral belief that an individual act ought to be (ought not to be, may be) performed. In other words, it might be thought that the tested premises are confirmed or disconfirmed merely by seeing whether or not they imply certain moral *beliefs,* and that one need not look any further. I hope that it is now clear that there *is* something beyond moral belief to which I think we may appeal, namely, the feelings of obligation and permissibility as I have explained them, just as there is something beyond descriptive belief to which we may appeal, namely, sensory experience. I do not subscribe to the coherence theory in my view of moral

8. Bertrand Russell, *The Problems of Philosophy* (New York, 1912), pp. 69-70, 72-73.

reasoning. Naturally, I think that all of our beliefs ought to be coherent or compatible but I do not think that this is a sufficient condition for accepting a conjunction of beliefs.

7. Ethical Argument and the Hypothetico-deductive Method

Since I have maintained that a so-called practical argument is a hypothetico-deductive argument, I want to say more about the role of hypothetico-deductive method in moral reasoning. Because I believe that any belief or statement which serves as a premise in a practical argument may be surrendered in the face of a recalcitrant feeling of obligation—and this is tantamount to saying that every such premise is advanced hypothetically and is therefore subject to revision or denial—I disagree with philosophers like Ross who think that true moral principles are self-evident; with philosophers who think that logical premises are analytic and therefore not rejectable by appealing to sensory experience, much less by appealing to emotion; and with philosophers who deny that the descriptive premises of a practical syllogism may be rejected as a consequence of someone's having a certain feeling. Such philosophers do not accept the view that *every* belief is accepted tentatively and that any belief may be surrendered under certain circumstances, no matter how firmly entrenched it may be. But even when this point is understood and accepted where all statements or beliefs are descriptive, there is a tendency to deny it when a moral argument is under consideration. *There is strong resistance to the idea that a moral argument may function as a hypothetico-deductive argument all of whose premises are subject to revision.* But once we relieve ourselves of this prejudice, thereby allowing that *no* premise is immune to revision, it becomes permissible to reject any premise in a moral argument if we should reject its moral conclusion. It should not be thought that we *cannot* use a practical argument as a hypothetico-deductive argument whereas we *can* regard the reasoning about the pharmacist's pills as hypothetico-deductive and corporatistic.

This prejudice is related, I think, to a legalistic view of ethical argument. For consider a situation in which a stern statute says that one who performs an act of a certain kind should be visited with capital punishment and where it is asserted that a person has performed an act of the kind in question. The logical conclusion will be that there is a legal obligation to visit capital punishment on the accused. In the view of some philosophers of law, the whole situation should be regarded as follows. The question whether the statute or major premise *is* a statute should be settled in one way, the question whether the accused performed a certain act is to be settled in another way, and the question whether the accused must be capitally punished should be settled merely by deduction from the statute and the statement of fact. So here we have a picture of legal thinking that is quite analogous to the picture against which epistemological corporatism is directed. The statute is viewed as something that is accepted because it has been handed down by a legislator. The descriptive statement that the accused has done what he is alleged to have done is thought to be of a different order and to be confirmed or disconfirmed by a fundamentally different kind of investigation. And, finally, the normative conclusion that the accused must be punished is thought to be necessitated by this conjunction of premises so that one can never, by challenging it, be led to reconsider any of the premises which have allegedly been established individually by techniques appropriate to each of them. Historically, this legalistic view of moral reasoning is kindred to the doctrine of natural law and Kantian ethics, neither of which is compatible with my version of corporatism. For each of these ethical doctrines has it that the major moral premise of a moral argument of the kind we have been considering is tested by a method entirely different from that used in testing the minor descriptive premise. And neither of these doctrines allows that *any* premise of a moral argument may be altered just because the conclusion is found unacceptable. Therefore, each of these moral doctrines is far removed from the view of ethical argument I advocate.

However, their sister-doctrine in jurisprudence may be challenged by those who hold a different view of law. I have in mind those who recognize that a jury's determination of the so-called facts in a case is often affected by its reaction to what follows from the conjunction of the statute and the prosecutor's description of the facts. Thus a jury which shrinks from the thought of electrocuting or hanging the accused will often issue a verdict on the ground that the accused did *not* commit murder if capital punishment is mandatory for murderers. Such a jury will be acting in a manner that is *analogous* to that of the corporatist in ethics because it would reject a descriptive statement in the wake of rejecting a normative conclusion. Of course, the jury is legally powerless to amend or reject the statute, and that is why it is limited to denying the prosecutor's description in a situation of this kind. But a wise legislator who observes juries behaving in this way might try to amend or repeal the statute. Like the jury, the legislator would be thinking along something like corporatistic lines. The jury would reject the prosecutor's description of the facts after reacting negatively to the normative conclusion of a legal syllogism whereas the legislator would, as it were, reject the statute in that syllogism after observing the negative reactions of juries toward that conclusion. Let me repeat, however, that I regard the legal illustration only as an analogy which I hope will clarify my version of corporatistic thinking. Those who do not find it clarifying are urged to forget it if they are persuaded that an ethical argument may function as a hypothetico-deductive argument whose descriptive premise or premises may be revised or rejected if the total body of premises leads to an unacceptable normative conclusion.

After mentioning the law in an effort to make a point about morals, I think it appropriate to quote a passage from Justice Holmes on morals, a passage that illustrates part of what I have in mind. Referring to "the Kantian injunction to regard every human being as an end in himself and not as a means", Holmes writes:

> I confess that I rebel at once. If we want conscripts, we march
> them up to the front with bayonets in their rear to die for a
> cause in which perhaps they do not believe. The enemy we treat
> not even as a means but as an obstacle to be abolished, if so it
> may be. I feel no pangs of conscience over either step, and natu-
> rally am slow to accept a theory that seems to be contradicted by
> practices that I approve.[9]

I do not quote this passage in order to express agreement with Holmes's substantive moral views but rather to call attention to a moral argument that helps illuminate what I have been saying. What Holmes has done here is ostensibly to test a single Kantian

9. Oliver Wendell Holmes, *Collected Legal Papers* (New York, 1920), p. 304.

moral principle, "Every human being ought to regard every human being as an end in himself and not as a means". He rejects this principle, but when we examine his method of rejecting it, we discover several things of interest. First of all, Holmes in effect attributes to the Kantian the following sort of argument:

(a) Every human being ought to regard every human being as an end in himself and not as a means.

(b) Each of these conscripts is a human being.

Therefore,

(c) Each of these conscripts ought not to be regarded as a means by any human being.

Secondly, Holmes rejects conclusion (c) because he *approves* of regarding a conscript as a means and, presumably, does not feel the obligation in question. Thirdly, Holmes moves immediately from his rejection of (c) to a rejection of (a). Of these three things that Holmes does, the second is closest to what a corporatist would do; it shows Holmes using his feelings in testing the conclusion of a moral argument. But if Holmes were a corporatist, he would consider the *possibility,* however repellent, of rejecting or amending (b) and "saving" the Kantian principle.

I do not say this in order to indicate that I myself would have actually taken this logical route but rather to indicate what might have been done by someone bent on weighing all of the relevant considerations that enter into thinking about such problems. Of course, my view will be denied by one who thinks that moral principles are tested by conscience or, alternatively, by using a device like Kant's supreme categorical imperative; that logical principles are tested by finding out whether they are, say, analytic; and that only so-called factual premises are to be tested empirically. In general, it will be denied by one who thinks that each premise in an argument of the kind under consideration is to be tested in isolation and, furthermore, that premises of different kinds are to be tested by different methods. It will also be denied by one who does not think that we may reject the conclusion of the above argument by appealing to feelings, even the feelings of normal persons under normal conditions.

Those who recoil from the idea that we may reject a descriptive premise because we have rejected the normative conclusion of an argument might ponder the implications of the following story. According to a newspaper report, a small American town decided to designate the first baby born to a town couple on January 1 of a centennial year as *the* town baby. On New Year's day Baby *X* was born at 5:15 A.M. and therefore was presumably the town baby of what we may call "Ourtown". But when it was learned that Baby *X* had been the child of unmarried parents, Baby *X* was said not to be the winner of the great honor. Instead, the honor was given to another child, born an hour later. A town official declared: "It was thought that a baby born to unmarried parents from Ourtown is not an Ourtown child". Note, however, that if the town's original commitment had been expressed by the statement: "Whatever Ourtown child is born earliest on January 1 ought to be identified as *the* town baby", and if the minor premise had been expressed by "Baby *X* was the Ourtown baby born earliest on January 1", it would obviously follow that Baby *X*, the child of unmarried parents, ought to be identified as *the* town baby. But once this normative conclusion was rejected, the town officials did not revise their major normative premise. Instead, they rejected the minor descriptive premise by flatly denying its implication that Baby *X* was an Ourtown child.

Of course, it might be said that the major premise in the illustration is not moral but legal. Yet the occasional inclination to deny that so-called white lies are not really lies illustrates the same procedure in morals. We might also take special note of the fact that the town fathers did not exercise the option of revising the major premise *ex post facto* by inserting the word "legitimate" between the words "whatever" and "Ourtown". After all, the rule had probably been indited in black and white somewhere, and for this reason it would have been inconvenient, impossible, or embarrassing to withdraw or revise the rule. This may be one reason why in more obviously moral illustrations the rule is likely to be preserved while the descriptive premises are manipulated. After all, moral rules have been traditionally viewed as God's commandments, written on stone and certified in Scripture. Therefore, *they* are not likely to be altered or rejected in situations of the kind we have been considering; instead, descriptive statements of fact usually provide casuists with more malleable candidates for amendment or alteration.

My use of the word "casuists" may stimulate certain readers to think that there is something *necessarily* wrong about the amendment or rejection of a descriptive statement after the fashion of Ourtown's officials. But it is certainly not my intention to use the word "casuist" in what the *OED* calls a "sinister sense"; on the contrary, my point is that if this is casuistry, it is not *necessarily* wrong to engage in it. It may be wrong on some occasions but not on others. I agree with Moore when he says: "The defects of casuistry are not defects of principle; no objections can be taken to its aim and object".[10] And although part of what I have said may be regarded as a defense of the *permissibility* of casuistry, I want to make clear that whether it is wrong or right to engage in casuistry on a given occasion depends to a great extent on whether its net effect on that occasion is to link sensory experiences and moral feelings as descriptive science links sensory experiences with each other. If we disapprove of the casuistry employed by the Ourtown officials, that is because we think that they replaced a good argument for the acceptable conclusion that Baby X *ought* to have received the honor by a bad argument for the unacceptable conclusion that Baby X *ought not* to have received the honor.

Because I have spoken so often of *arguments* in descriptive science and ethics, I want to say something about this mode of speech since my use of it may puzzle some readers. An argument is often thought of as a piece of reasoning which is presented in order to convince someone of the truth of a certain statement. In that case it might also be thought that the arguer first makes statements of whose truth he is sure and then deduces other statements until he reaches the truth to be proven. Indeed, this conception of a demonstrative argument lies behind the idea that any conclusion of an argument must be accepted as certainly true. However, a piece of deductive reasoning that may be called an argument need not begin with premises which the arguer accepts as certainly true. On the contrary, he may merely wish to see whether he *should* accept them as true, and with this in mind he deduces a consequence from them that he can test by appealing to experience. In this case we regard his argument as hypothetico-

10. *Principia Ethica*, p. 5.

deductive. His premises are put forward tentatively in order to see whether they will be confirmed by implying a conclusion which will in turn be confirmed by data. However, he is also prepared for recalcitrant data that disconfirm the conjunction of the argument's premises.

What may be confusing about the situation is that an arguer may put forward a piece of reasoning with full confidence that he is arguing demonstratively—that is to say, from premises of whose truth he is sure—and then, by the time he has deduced the conclusion, find that the latter is rebuffed by the data. If he then looks back with a quizzical eye at his premises, the argument that started out in his mind as demonstrative is now viewed differently by him. This lies at the heart of the idea that descriptive scientists should always advance their premises as hypotheses which are collectively tested by deducing consequences that are to be checked. And what I have done is to extend this idea so that it covers moral reasoning, while emphasizing that the rejection of a normative conclusion may affect even a descriptive premise. Of course, one may have great confidence in a moral principle that one uses as a premise in what one may initially view as a demonstrative moral argument, just as one may have great confidence in a theory that one uses as a premise in what one may initially view as a demonstrative physical argument. Yet one must realize that this initial confidence falls short of the self-evidence that some philosophers have attributed to moral principles. The same thing may be said about the descriptive premises used in moral arguments. They too should be viewed as hypotheses that may be rejected or altered after checking the consequences of conjunctive statements.

Because I have expressed disagreement with Kant earlier in this section, I want to call attention to a passage in which he explicitly rejects the epistemology of moral principles that I have been defending. The passage occurs in his essay, "On a Supposed Right To Lie From Altruistic Motives" ("Über ein vermeintes Recht aus Menschenliebe zu lügen"), where he argues against Benjamin Constant that there is no such right. However, at one point Kant approvingly quotes Constant, whom he calls "the author", as follows: "The author says, 'A principle recognized as true (I add, recognized as an a priori and hence apodictic principle) must never be abandoned' ".[11] This is the view which I reject when I say that we may use the hypothetico-deductive method in ethics and when I say that we may alter a conjunction of premises in a moral argument by abandoning a *so-called* "a priori and hence apodictic principle" by appealing to emotion.

It should be evident by now that the emotion or feeling to which we appeal in moral argument is not any old emotion, as it were. Not only is it an emotion which is felt by a certain kind of person under certain circumstances, but because it is an emotion or feeling that is peculiarly relevant to judgments of duty and right, it will be a feeling of obligation or of right directed toward an act that happens to be of a certain kind. Consequently, the feeling is not one of rage, boredom, uneasiness, etc. When we are evaluating the premises of a moral argument by appealing to emotion and experience, the range of relevant emotion is narrower than the range of sensory experience. So narrow is it that a feeling of obligation is not even to be identified with a feeling of *approval*. Perhaps every feeling of obligation is also a feeling of approval, but the converse is not true. For this reason, it will not be enough to say, in confirmation of a belief that we are obligated to perform a certain act, that we have a feeling of approval toward the act. It will therefore not do to say that statement (8) is a conjunct in a conjunction such as the following: "The prisoner's act of saying yesterday at 4 P.M. 'My regiment went north' is an act that ought not to have been performed; *and* I am a normal person; *and* I am contemplating the prisoner's mendacious utterance under circumstances that correspond to viewing the powder under white light; *and* when, under such circumstances, a normal person contemplates an utterance which ought not to be made, that person will *disapprove* of performing the action". The feeling about the act is the feeling of its being *required* and not the feeling of its being *approved*. The former, not the latter, is the feeling that is to be linked with sensory experiences when we are testing a mixed conjunction of premises. However, if the feeling of approbation *were* the feeling to be linked, the claims of limited corporatism would still stand. Corporatism does not of itself dictate what feelings are to be linked with experiences.

11. Kant, *op. cit.*, p. 429. Once again I use a translation that appears in L. W. Beck's edition of a volume of Kant's writings, *Critique of Practical Reason, and Other Writings in Moral Philosophy*, p. 350.

8. *Less Direct Ways of Rejecting a Normative Conclusion*

Lest it be thought that I espouse an excessively simple view of the way in which recalcitrant emotion may lead us to reject a moral conclusion, I want to say something about ways that are more indirect. I have so far concentrated on the comparatively direct impact of a discordant feeling or emotion on such a conclusion, but one may show by a more complicated method that the prisoner mentioned in that illustrative statement was not obliged to refrain from doing what he did. I want now to offer some examples of more complicated and less direct ways of rejecting the singular moral conclusion about the prisoner's action while continuing to accept the view that one must reckon with emotion, albeit indirectly.

Since the original conclusion was that the prisoner *ought not* to have said mendaciously that his regiment went north, one way of rebutting it is to show *by argument from other principles* that for other reasons the prisoner *ought* to have said that his regiment went north. For example:

(9) Every act that leads to saving the lives of one's countrymen is an act that ought to be performed.

(10) The prisoner's act of saying yesterday at 4 P.M. "My regiment went north" is an act that led to saving the lives of his countrymen.

Therefore,

(11) The prisoner's act of saying yesterday at 4 P.M. "My regiment went north" is an act that ought to have been performed.

In this argument we come to a conclusion which forces us to give up (8) unless we abandon other beliefs. A prisoner who was trying to decide whether to lie could not accept both conclusions. But if we do give up (8) in this way, we no longer assert (6) and (7) since we now approach the prisoner's act in an entirely different manner. We use another moral principle and we describe the prisoner's act differently. We might say that when "looked at" in the way represented by (6), (7), and (8), the prisoner's act was one that he ought *not* to have performed whereas, when looked at in the way represented by (9), (10), and (11), it is one that he *ought* to have performed. We may take the latter logical approach if we can defend a way of ranking principle (9) above principle (6), arguing that when an act is one of lying *and* is also a patriotic act, our moral judgment of it should nevertheless be governed by considering it solely as an act of lying because we regard the duty not to lie as higher than that of patriotism. But in that case our decision to appeal to the supposedly higher duty will in effect dictate what we regard as *the* feature of the act for purposes of moral judgment. Even though we know that the soldier's utterance is mendacious, we may disregard that feature of it in order to focus on its being a patriotic act. For the answer to the question "What did the soldier do?" may vary, just as the answer to the question "What is that thing?" may vary, depending on the standpoint from which we view the act.

We often classify one and the same act in different ways, that is to say, describe it in different ways, depending upon the office or role we happen to occupy. William James says that we discriminate among what he calls our different selves.[12] And so, depending upon what role we regard ourselves as occupying at a given time, we may consider our prisoner's utterance merely as a lie or merely as a patriotic act. If we consider it in the first way, we may come to conclusion (8), whereas if we shift selves, as it were, and consider it in the second way, we may come to conclusion (11) and hence abandon (8). If one person happened to be both a soldier and a moral philosopher, he might well "see" the prisoner's act in different ways, depending on which of his Jamesian selves he found himself occupying at a given moment. As a military man he might regard the utterance merely as a patriotic act and subsume it under principle (9), but as a philosopher he might regard it merely as a lie and subsume it under (6). The problem is more acute, of course, for a philosopher-prisoner who is trying to decide whether he ought or ought not to lie in the middle of a battle. Will he not have to rank these different selves and their respective viewpoints if he wants to come to a conclusion about whether he ought or ought not to lie? Such a ranking would be similar to what I have called the ranking of moral principles and would itself be moral in character. It would of

12. William James, *The Principle of Psychology* (New York, 1890), Volume I, pp. 294ff.

course be extremely difficult to carry out in a systematic way and it should be noted that some principles may be ranked equally. Here we find the source of what is often regarded as a tragic situation.

In light of all this, we can see how hitherto unmentioned moral considerations may enter our thinking about so-called facts. Since a moral judgment underlies our decision that one moral principle rather than another which it outranks should preside as the major premise of our practical argument, this moral judgment indirectly determines how we shall *describe* the judged act in our minor premise. And in the wake of this we may set up the new practical argument consisting of (9), (10), and (11) in place of the old one consisting of (6), (7), and (8). If we put the prisoner's act into a cubby-hole that is controlled by a supposedly higher-ranking moral principle, we are led to the conclusion that is implied by a new pair of premises. As an act of lying, the prisoner's act *ought not* to have been performed but, as an act of saving his countrymen's lives, the soldier's act *ought* to have been performed. In selecting the latter cubby-hole after reflecting on the prisoner's action, we indirectly reject our original moral conclusion.

Although this way of rejecting our original moral conclusion is more indirect than the one we use when we appeal directly to a *feeling* about the man's not being obligated to refrain from an act described as one of lying, we do not cease to appeal to feeling when we take the indirect route. For feeling plays a part in the acceptance of our new pair of premises. The prisoner's act engenders feelings as well as sensory experiences that must be linked by the new pair of premises if we are to accept it; and a decision to put this pair in place of the old pair depends on a belief that the new pair more successfully links what we experience and feel than the original pair did.

So far we have considered only one mode of indirectly rebutting a singular moral conclusion such as (8), but there is at least one other mode of indirect rebuttal that is also worth our attention. We may hold that although we are bound not to lie, an exception to the rule may be made when lying is dictated by the supposed obligation to save the lives of our countrymen. Under such circumstances, we might set forth an argument such as the following:

(12) Every act which is an act of lying but one that leads to saving the lives of one's countrymen is an act that may be performed, i.e., that we have a right to perform.

(13) The prisoner's act of saying yesterday at 4 P.M. "My regiment went north" is an act of lying but one that leads to saving the lives of his countrymen.

Therefore,

(14) The prisoner's act of saying yesterday at 4 P.M. "My regiment went north" is an act that he had a right to perform.

In asserting (12), we make an exception to the moral principle that forbids lying and we do not imply that the prisoner has an *obligation* but rather that he has a *right* to lie. This means that we would not condemn a prisoner who refrained from lying, as we would if we had accepted, instead of (12), the principle "Every act which is an act of lying and one that leads to saving the lives of one's countrymen is an act that *ought* to be performed".

Now that we have considered two indirect ways in which we can reject normative conclusion (8) as derived from (6) and (7), let us compare these ways. We have seen that when we reject (8) after deducing (11) from (9) and (10), one result is not the *denial* of descriptive statement (7) but rather its *disappearance* from our moral reasoning. We exchange one argument for another that is entirely different and we cease paying attention to the fact that the prisoner lied. But the situation is different when we reject (8) by arguing for (14). In that case we do pay attention to the fact that he lied but that is not all that we assert about what he did; we add that his act led to saving the lives of his countrymen and thereby *amend* (7). However, whether we change our premises completely or merely amend them, the change in our descriptive premise is in each case linked with moral considerations. We view the prisoner's act as an act of lying without qualification in statement (7), as an act of patriotism without qualification in statement (10), or as an act of lying patriotically in statement (13), depending on what moral principle we think should preside over the argument that leads to our singular moral conclusion.

However, the new conjunction of premises, whether it be totally or partially different from the original conjunction, must, if it is to be preferred, organize the moral judge's sensory experiences and moral emotions more effectively than the conjunction it replaces. And when we put one moral principle above another, I think we do so because we think that the out-ranking principle will, when it is conjoined with a relevant description of the act, generally link experience and moral feeling more effectively than the out-ranked principle will. Thus, if we set the principle that forbids lying above the principle of patriotism, that is because we think that arguments like that from (6) and (7) to (8) tend to organize our experiences and emotions more successfully than do arguments like that from (9) and (10) to (11).

The corporatistic idea that moral principles are ranked and tested by their capacity, *in conjunction with descriptive statements,* to organize sensory experience-cum-moral emotion is different from the views of Kantians, of some utilitarians, and of advocates of the doctrine of natural law in a fundamental respect. All of these philosophers think that we can show whether a moral principle is acceptable by subjecting it to a test which requires us to regard it as an isolated statement. The Kantian tells us to test it merely by asking whether any rational being could consistently choose that it be accepted and acted on by everyone; some utilitarians tell us to test it by asking whether accepting it and acting on it maximizes pleasure; and the advocate of natural law tells us to test it by asking whether it is self-evident or deducible from a self-evident truth. All of these philosophers are therefore committed to examining a moral principle by itself and not as a principle which works conjointly with descriptive statements in moral arguments. All of them think that we can, *without discovering how a moral principle works in concrete moral arguments,* certify it as acceptable. By contrast, I reject this approach for the same reason that I would reject the idea that we should test the descriptive principles of natural science without discovering how *they* work in their natural corporate habitat.

9. *Bodies of Belief*
Linking Sensory Experiences and Moral Feelings

I now wish to expand on my view that heterogeneous bodies of belief which contain both normative and descriptive beliefs ought to link sensory experiences and moral feelings, a view that I have distinguished from the view that bodies containing only descriptive beliefs ought to do this for sensory experience alone. I have advocated an enlargement of our conception of a body of beliefs so that it may in certain instances include normative beliefs as well as descriptive beliefs, and I have also advocated a corresponding enlargement of the data to be organized by such an enlarged body of beliefs. I cannot see how any philosopher could successfully object to this double enlargement by asserting that we have no normative beliefs or by asserting that we have no moral feelings.

So far I have concentrated on defending that element of my version of corporatism which concerns the mixed body of descriptive and normative beliefs. But how is the combination of experience and emotion of which I have spoken related to the mixed body or conjunction of premises of a given hypothetico-deductive argument whose conclusion is a normative singular statement? One view that I cannot accept is the view that we may test a descriptive premise of the mixed argument *in isolation* by consulting experience, and that we may test the normative premise of that argument *in isolation* by consulting moral feelings. On the other hand, a descriptive premise in a mixed argument, for example, statement (7) that the prisoner lied, might have been tested corporatistically along with other descriptive statements in some other totally descriptive argument; and, on the basis of that prior test, it might have been asserted as a premise in the mixed argument for (8). Similarly, (6), the moral premise of that mixed argument, might have been tested corporatistically in some other mixed argument; and on the basis of that prior test, *it* might have been asserted as a premise in the mixed argument for (8). But even though (6) and (7) have prior corporatistic support, when (6) and (7) are conjoined in this new argument for (8), they are put in jeopardy once again simply because (8), their conclusion, may face a recalcitrant emotion or more indirect challenges.

Let me suppose, then, that (8) *does* face a recalcitrant feeling of a normal person under normal circumstances and that the person who has that feeling decides to reject (8). If he then rejects descriptive statement (7), he does something that some philosophers do not regard as permissible. According to them, he rejects a descriptive statement which has, along with other descriptive statements, linked sensory experiences by the standards of corporatism but, they say in criticism, he rejects it on the basis of feeling. He will therefore be pitting his feeling that the prisoner was not under an obligation not to lie against a descriptive belief which should be—according to some philosophers—impervious to this kind of opposition. And yet I maintain that he *may* do this within the bounds of rational thinking because he rightly believes that his task is to organize or link both his sensory experiences *and* his moral emotions in an acceptable manner.

I repeat, however, that I do not hold that a purely descriptive body of physical statements accomplishes that sort of linking or organization. A physicist who confines himself to making statements that do not contain normative words need not do this in order to justify his bodies of statements. But a moralist ought to link sensory experience and moral emotion. If he erects a linguistic framework in which some sentences are purely descriptive and others linguistically mixed in the sense in which moral principles are mixed, he erects that kind of framework because he is trying to organize or link more than his sensory experiences. When I say that a moral principle is mixed, I mean that it says that all acts to which a descriptive term is applied are acts to which a moral term is then applied. Such a mixed sentence is what gets one from the descriptive minor premise of a practical argument to its moral conclusion. In such an argument it is the moral principle that makes the conjunction of premises mixed since the moral principle must be supplemented by a descriptive premise if one is to derive a normative conclusion. Moreover, as soon as one thinks of a moral principle as a mixed sentence which one cannot establish individually on the basis of sensory experience alone, one is on the way to seeing more clearly that the presence of such a moral principle not only turns the whole body of belief into a mixed body but shows that the believer is trying to organize his sensory experiences and moral feelings.

This does not mean that a practical argument which is being tested must be given two separate marks—one for its capacity to organize sensory experience and the other for its capacity to organize emotion—before we may render a final judgment about the corporate body's capacity to organize sensory experience and emotion. The rendering of separate grades before issuing a final judgment here would be as unnecessary as it would be in the case

of purely descriptive bodies of belief that are supposed to organize only experiences associated with the normal five senses. In
the latter case one would not be expected to answer five different
questions like "To what degree does the body of belief organize
visual experience?", "To what degree does it organize auditory
experience?", "To what degree does it organize olfactory experience?", and so on. The final judgment about a body's capacity to
link or organize different kinds of sensory experience does not
depend on five such answers. And, by the same token, the corresponding final judgment about a heterogeneous or mixed body's
capacity to link sensory experiences and a feeling of obligation
does not depend on two answers that are derived by separately
evaluating the body's power to organize sensory experience and
its power to organize emotion. Rather it depends on whether the
body of beliefs can get us, so to speak, from certain sensory experiences to a certain moral feeling in an acceptable way.

My use of the phrase "get us" suggests that it is helpful to
think of a mixed body of beliefs as a bridge of several parts that
gets us from such experiences to such feelings. In our example
about abortion we have sensory experiences that are linked by
the argument that contains the statement that the mother killed
the fetus, the statement that every living fetus is a human being,
and a logical principle. And these experiences might be appealed
to in corporatistic arguments for each of these premises. We also
have a feeling that is especially relevant to the belief that in committing an act of abortion the mother did what she ought not to
have done. Since the purpose of building a bridge that contains
these different parts which represent different kinds of statements
or beliefs is to link these experiences and this moral feeling, we
may let the first part of our figurative bridge represent statements
by which our sensory experiences are linked while letting the
third part represent that statement to which our feeling is especially relevant—the conclusion. The second part, the middle of
the span, represents the moral principle. Now when we move
across the bridge from our experiences to our moral feeling without any hitch, we are not led to examine these parts very care-
fully, as William James and others have emphasized. While
moving smoothly, we do not think we need any repairs in the
structure. But let us suppose that the bridge no longer takes
us smoothly from sensory experiences that we have used in

corporatistically supporting the descriptive premises to moral feelings that we have used in corporatistically supporting the conclusion. Suppose, in other words, that when we get to the other side, the kind of moral feeling we reached on previous trips is no longer ours. Indeed, suppose that our new feeling runs counter to the feeling to which the bridge has led us in the past. And suppose, too, that we are moved by this to repair the bridge and that we begin by replacing the third part, the belief that the mother who killed the living fetus did something that she ought not to have done, by an opposed belief. We thereby begin to accommodate our present feeling in the corporatistic manner I have sketched. But so long as we hold on to our logical principle, we see that our bridge is so constructed that we cannot be content with changing the third part of our bridge. We have to make other changes. Where shall we make them? If we hold on to our logical principle, then we *may* make them in the middle of the span *or* we may make them in the first part of the span. Either one of these changes is *permissible* in my view in spite of the protest of some philosophers who think we may make changes only in the middle of the span after making them in the third part.

Once either sort of change is made, our bridge will link experiences and moral feeling more effectively because it will *get us* from our sensory experiences to our relevant moral feeling. I should point out that even if we deny or amend the principle that a living fetus is a human being in our overhauling of the bridge, we need not deny having the experiences that we thought supported that principle. If physicists may plead parallax, hallucination, or poor eyesight when they do not let their acknowledged sensory experiences dictate a belief, moralists may do something similar with regard to the first or descriptive part of their bridge. They may sincerely plead analogous failings after they have said that the mother killed a fetus, thereby accommodating their feeling that she did not do something she ought not to have done. Or they may take care of all of the experiences that were thought to have supported "Every living fetus is a human being" while denying this statement. All of this may seem obvious when the accommodating is done in response to a recalcitrant sensory experience but it seems objectionable to some philosophers when it is prompted by a recalcitrant moral feeling. My point, however, is that we may, even in the latter case, make alterations in the first part of our bridge because the purpose of building the bridge is to link experience and moral feeling in a rational way by means of a logically organized linguistic structure. Someone may object that our new bridge does *not* link our experiences and our moral feelings, but *that* is a different objection. It is compatible with accepting *the right* to repair the first part of our bridge in compensation, so to speak, for our repair of the third part when the repair of the third part has been prompted by a discordant moral feeling.

10. *Dealing with Some Other Objections*

I want now to make a further effort to meet possible criticisms. I am well aware that one fundamental objection to my view is that it allows so-called facts to be twisted in the interest of feeling, which allegedly should be kept out of objective inquiry. Those who protest in this manner might say, as I have indicated, that if a practical argument is treated as a hypothetico-deductive argument and if its conclusion should be rejected by someone, the only step that such a person could take if he were not disposed to challenge the logical principle employed in the argument would be to reject or amend the moral principle in the argument. But why does the protester seize on the moral principle as the only vulnerable statement in such a situation? Because, as we have seen, the protester may insist that if you reject a *moral* conclusion, then the only kind of premise you are entitled to alter is a moral one. Aside from denying this for reasons that I have already presented, I think it misses an important similarity between what a moralist can do in such circumstances and what a physicist can do in analogous circumstances. The moralist's decision to give up a descriptive minor premise rather than a long-accepted moral principle is not fundamentally different from the physicist's willingness to assert the existence of novel forms of energy or novel particles in order to "save" the vaunted principle of conservation of energy. Why, then, shouldn't the moralist alter his body of beliefs as the physicist alters his if the moralist does so because he aims at a comparatively simple organization of his data which does not disturb his body of beliefs excessively as he makes his alterations? I realize that when we "save" a moral principle and give up a descriptive statement under circumstances like those I have described, we may give the impression of *misrepresenting* "the facts". But we can see the similarity between what we do here and what we do in the corresponding situation in physics *if* we regard a practical argument as a hypothetico-deductive argument which is used by a moralist to link sensory experience and emotion while the moralist, like a physicist, respects older truths and values scientific simplicity.

This respect for older truths and this search for simplicity constrain believers in all domains of rational inquiry. These constraints are formulated in epistemic normative maxims that govern the process of establishing and disestablishing beliefs, maxims that go beyond Duhem's maxim that one should exercise good sense and employ Pascalian "reasons which reason does not know" in trying to decide which hypothesis to abandon in an intellectual crisis.[13] One of these, I think, is a maxim that would counsel the believer to exercise caution in exercising the right to surrender descriptive statements when dealing with bodies composed of descriptive statements and moral statements, that is to say, with bodies of the kind discussed so far. This should mollify some critics of the view I have been advancing, for it would acknowledge the dangers of recklessly amending or surrendering descriptive statements in the manner I have described. Needless to say, however, such critics should realize that our sensory experiences are subjective insofar as they are *ours* and that the idea that a purely descriptive body of statements should rationally link those experiences with each other might also lead us to worry about our objectivity in some sense of that difficult word. The corporatistic testing of a statement like "This powder is white" in conjunction with a statement about normality of vision, with a statement about the whiteness of the light in which the powder is viewed; and with the generalization that whenever a normal person looks at a white thing in white light it will appear white to him is devised in order to remove the taint of subjectivity or idiosyncrasy that is likely to be associated with all testing by reference to the sensory experience of a human being. But, as I have already indicated, an analogous corporatistic testing of singular statements about obligations by appealing to what

13. Duhem, *La Théorie physique*, p. 330.

normal persons would feel under normal conditions is intended
to remove the taint of subjectivity or idiosyncrasy associated with
all testing by reference to emotion. I do not doubt that we can
get more consensus about the whiteness of powder than we can
get about the obligatoriness of an act, but I cannot accept the
notion that sensing a thing somehow gets us closer to its "real"
color than feeling gets us to its "real" moral quality if our feeling
is supposed to be that of a normal person in normal circumstances just as our sensing is. True, it may be harder to get people to agree about what normality is in the case of feeling, but
the difference in question is one of degree that reflects what everyone admits—namely, that agreement in descriptive science is more
common than agreement in morals. We cannot *prove* that sensing gets us *closer* to the "real color" than feeling gets us to the
"real moral quality".

Hume denied that so-called moral qualities really belong to
the actions to which we ordinarily attribute them. He thought,
to use his own figure, that taste gilds or stains all natural objects
with the colors borrowed from internal sentiment whereas reason discovers objects as they really are in nature. But I deny that
we can make the implied distinction between predicates such as
"white" and predicates such as "obligatory". Indeed, Hume himself says that vice and virtue may be compared to sounds, colors,
heat and cold, "which, according to modern philosophy, are not
qualities in the objects, but perceptions in the mind". And he
goes on to say in a passage to which I shall return in Chapter 6
that the view he advocated in morals, like the above one concerning sounds, colors, etc., "is to be regarded as a considerable
advancement of the speculative sciences".[14]

People may more readily converge in their judgment that an
object is white than they may converge in their judgment that an
action is obligatory, but this does not mean that those who agree
that an object is white do so because of the object's real whiteness
whereas those who disagree about whether an act is obligatory
disagree because the act *lacks* a moral quality. To be sure, some
philosophers cite the variation of moral opinion in defense of the
view that moral qualities are not real qualities of actions, but in
my opinion they do not do so successfully. I think that both

14. Hume, *A Treatise of Human Nature*, p. 469.

"obligatory" and "white" express objective or real properties, and I explain the variation in moral opinion by pointing out that it is difficult to agree about the application of predicates like "obligatory". Furthermore, there are hosts of predicates which are thought to express objective properties but which are just as hard to apply with the agreement of others. Consider the word "sensitive" when used in ordinary English about a person. Many think that it is an objective psychological term, but it is very hard to get people to agree that a person is or is not sensitive. The denial that a term is objective cannot be justified merely on the basis of our disagreements or difficulties in applying it. The word "true" is an objective word for most philosophers and yet think how much disagreement there is about whether certain statements are true.[15]

15. Some things that I have said in this chapter and in part IV of my *Toward Reunion in Philosophy* are in accord with certain remarks about moral theory that John Rawls makes in *A Theory of Justice* (Oxford, 1971), pp. 46-53. I have in mind his view that "analyses of meaning" do not have a special place in moral theory (I would say they have no place); that moral philosophy must be free to use contingent assumptions; and that a moral theory is concerned with the moral sentiments.

IV

Corporatism, Metaphysics, and Epistemology

11. *How Our Metaphysics May Be Affected by Our Normative Beliefs*

Although I have so far concentrated on the ways in which the rejection of a singular normative belief may lead to the rejection or amendment of a singular descriptive belief of psychology or history, other sorts of descriptive beliefs may be rejected or amended because we have rejected a universal normative belief. For example, a universal metaphysical thesis such as determinism might be treated in this way after someone has rejected the universal conclusion of the following argument that runs from (15) through (19):

(15) Every physical event is necessitated.
(16) Every human action is a physical event.

Therefore,

(17) Every human action is necessitated.

So far we have made only descriptive statements since none of them is normative; none of them says what ought to be done, what ought not to be done, or what may be done. But now we *do* make a normative moral statement:

(18) If a human action is necessitated, it ought not to be judged morally.

From the above premises we then deduce another normative moral statement:

(19) No human action ought to be judged morally.

And finally we deny (19), just as we might deny the singular moral conclusion of a previously considered moral argument.

The main point I should like to make is analogous to the point I have made about practical arguments having singular conclusions. After rejecting (19) one might choose to reject (15), the descriptive thesis of determinism; one might reject (16); or one might instead reject moral statement (18). Therefore, we *need not,* as some philosophers seem to think, reject determinism alone if we reject moral statement (19). However, we *may* reject determinism in this instance.

I am not defending the argument I have outlined; I am merely trying to show that it can plausibly be made. In certain respects it is similar to the other arguments I have used in illustrating my claim that a descriptive statement may be surrendered in order to accommodate the denial of a normative conclusion. However, it is also different from those other arguments insofar as the rejected descriptive thesis of determinism comes from metaphysics, unlike the psychological descriptive statement "The prisoner's act of saying yesterday at 4 P.M. 'My regiment went north' was a lie". Therefore, the corporatism I have advocated not only permits our normative beliefs to affect the descriptive, that is to say, non-normative beliefs we may hold in natural science but it also permits our normative beliefs to affect the descriptive beliefs we may hold in metaphysics.

As before, however, I want to emphasize the right to reject *any* of the premises after having rejected the conclusion. And I want to point out that a philosopher who holds that all of the premises in this most recent illustration are self-evident, analytic, a priori, or logically necessary, might be inclined to conclude that none of them may be rejected in the manner I describe. If he does, then he will think of conclusion (19) as inescapable. He will cut off any effort to reject any of the premises on the basis of rejecting the conclusion simply because he will regard the premises as certain and undeniable. He will say that since all of the premises are undeniable, the argument cannot be regarded as hypothetico-deductive and that the conclusion has been *demon-*

strated because it follows inescapably from unassailable premises. Such a view of the situation will be taken by a philosopher who holds that not only *descriptive* premises such as (15) and (16) are unassailable but also that a moral premise like (18) is.

It should be noted that (18) is closely related to the Kantian dictum that "ought" implies "can", which I interpret as a moral statement without ascribing this interpretation to Kant himself. I do not regard the Kantian dictum as a statement of logical implication because I do not think that we use the word "implies" in it as we use the word "implies" in the statement that "man" implies "animal" or in the statement that "square" implies "rectangle". Analogously, I do not think that "necessitated" logically implies "ought not to be judged morally". Certainly the statement "Whatever is necessitated ought not to be judged morally" is not analytic by Kant's own standards, since it is hard to see how even the most confident employer of the idea of conceptual containment could say that the concept of being necessitated contains the concept of being an act that ought not to be judged morally just as *man* contains *animal*. Rather, I think that one who subscribes to this alleged implication wants to say what is said in statement (18); he is making a moral judgment to the effect that a certain kind of action ought not to be judged morally.

It is clear that we do make moral judgments *about* making moral judgments. Thus, the New Testament at one place contains the words "Judge not, that ye be not judged" (Matthew 5:48); and at another "Judge not, and ye shall not be judged" (Luke 6:37). It makes perfectly good sense to say that a person ought or ought not make a moral judgment to the effect that someone ought or ought not to perform a certain action. What has been doubted, however, is whether statement (18) presents a sufficient condition for attributing a moral obligation to refrain from judging an action *morally*. For example, Isaiah Berlin[1]

1. I have expressed views like those advanced here, as well as related views, in a number of other places. See my review of Isaiah Berlin, *Historical Inevitability* (London, 1954), in *Perspectives* USA 16 (Summer, 1956), 191-96, reprinted in my *Religion, Politics, and the Higher Learning* (Cambridge, Mass., 1959), pp. 75-84; and my *Foundations of Historical Knowledge* (New York, 1965), Chapter 7 *passim*. Berlin has responded to my criticisms of his views on this subject in his *Four Essays on Liberty* (London, 1969), pp. xix-xxiii, xxxvi-xxxvii. I have replied to the latter in my "Oughts and Cans", in *The Idea of Freedom: Essays in Honour of Isaiah Berlin*, ed. Alan Ryan (Oxford, 1979), pp. 209-19.

criticizes my view and holds that when we say that no caused or necessitated action should be judged morally, we do not, as I think, make a second-level moral statement about moral judgment. Rather, Berlin holds, we are saying that a sentence by means of which we purport to assert that a necessitated action ought (ought not) to be performed is in some sense incomprehensible. Therefore, for Berlin, the sentence "His choosing to steal the bracelet was caused but he ought not to have stolen the bracelet" is incomprehensible. I cannot accept Berlin's view for reasons I have tried to present elsewhere.[2] In my view, when we say that no necessitated action is blameworthy (or praiseworthy), we are asserting a moral proposition. If I am correct about this, then even if a philosopher should assert that premises (15) and (16) are necessary or analytic, the fact that premise (18) is a moral principle would prevent such a philosopher from maintaining that each statement in the argument, including premises and conclusion, is necessary or analytic. Because I do not regard moral principles as necessary or as analytic, I do not think that the reasoning from (15) through (19) establishes the necessity or analyticity of (19). On the contrary, since I regard premise (18) as a moral statement and therefore also regard (19) as a moral statement, I believe that one who rejects moral statement (19) may reject any one of the premises that lead to it, including the non-moral premises. Moreover, one who rejects (19) might do so, in accordance with corporatism, by selecting a particularly horrible deed, such as Hitler's treatment of the Jews, in order to show that if (19) were accepted, we would be obliged not to judge Hitler's act morally. Therefore, if we insisted that Hitler's act *ought* to have been morally judged, we could reject (19) and from there move to the rejection of the principle of determinism as expressed in (15), or any of the other premises, as I have indicated.

On the other hand, after denying (19), we might engage in some interpretation that would prevent the deduction of (19) even though we might be prepared to accept all of the statements from (15) through (18). Such an interpretation would focus on the word "necessitated" and would constitute a familiar move in the history of discussions of free will; it would consist in saying

2. See my "Oughts and Cans", cited in note 1 above.

that the allegedly elliptical word "necessitated" in (15), the principle of determinism, should be expanded differently from the
way in which it should be expanded in (18), the moral rule that
forbids judging a necessitated action. In (15), as this familiar
move would have it, "necessitated" is elliptical for "necessitated
by something", whereas in (18), it is elliptical for "necessitated
by duress". If, therefore, we expand statement (15) into "Every
physical event is *necessitated by something*" and then expand
statement (18) into "If a human action is *necessitated by duress,*
it ought not to be judged morally", we cannot deduce statement
(19). For, in keeping with these expansions, (17) will have to be
expanded into "Every human action is necessitated by something", a statement which, when conjoined with "If a human
action is necessitated by duress, it ought not to be judged morally", will not yield (19) as it stands in the original argument.
The upshot of these expansions is to save determinism after denying that no human action ought to be judged morally.

I wish to stress, however, that it was the rejection of moral
statement (19) that led to these alterations of our original premises, alterations that resemble an alteration we have previously
discussed. I have in mind the proposed rejection of statement (3),
"Every living fetus in the womb of a human being is a human
being", in response to the rejection of statement (5) that the
mother who killed the fetus thereby did something that ought
not to be done. It might be argued that the proposed rejection
of (3) involves a reinterpretation of the expression "human being" in such a way as to exclude a fetus from the class of human
beings. And this is similar to our treatment of the word "necessitated" insofar as it too involves reinterpreting language in order to avoid a certain objectionable moral conclusion. An important difference, however, between the two reinterpretations
derives from the fact that the more recent one concerns the morality of making moral judgments whereas the earlier one concerns the morality of taking lives. Apart from that, they are quite
similar.

It is important to stress this similarity because it might be
thought that the debate over free will is confined to the realm of
descriptive statements and therefore does not involve any substantive moral beliefs about what may be morally judged. But if
my approach is correct as outlined, the debate forces us to refer

to our moral views about morally judging certain kinds of actions. We must acknowledge that the debate over free will that appeals to the dictum " 'Ought' implies 'can' " is hardly "value-free". The philosopher who rejects determinism in order to avoid the conclusion that no act ought to be judged morally is moved in part by his moral beliefs about moral judgment and so is the philosopher who tries to save determinism by making distinctions between the necessitation expressed in (15) and that expressed in (18). Both of them want to have the right to pass moral judgment on *some* actions even though they differ about the metaphysical price they are willing to pay for this right. One philosopher is prepared to give up determinism to make those moral judgments whereas the other wants to hold on to determinism *and* to make those judgments. If the latter did not think he was obliged to make moral judgments and therefore did not deny conclusion (19), he might not take the trouble to make a distinction between the necessitation expressed in (15) and that expressed in (18). Therefore, it is evident that some participants in the debate over free will are not interested only in the logical analysis of necessitation or in the defense or rejection of determinism; they are also concerned with a moral issue about the right to make moral judgments. And because their acceptance or rejection of determinism, their acceptance or rejection of what I called Kant's dictum, and their acceptance or rejection of the statement that no human action ought to be judged morally, are all linked, they are brought to the point where they must decide which combination of these acceptances and rejections they wish to advocate. In order to decide, they should abandon the idea that only one such combination may be defended in a statement-by-statement investigation which is carried out by consulting one's intuition or by peering at the world of concepts. Instead, they should examine each alternative combination of beliefs in order to see which of them links experience and moral feeling more successfully.

Before concluding this section I want to deal with one more matter. Although I have emphasized that statement (18) is a moral statement while trying to show that determinism might be surrendered by someone who rejected moral conclusion (19), I do not hold that a rejection of determinism could not be carried out if the connection between being a necessitated action and being one that ought not to be judged morally were *not* moral. For even if a statement of the connection between being a necessitated act and being one that ought not to be judged morally were *not* moral but rather what might be *called* analytic or necessary, the statement *could* be rejected. I have made that clear, I hope, at many earlier places in this discussion. Why, then, do I emphasize that (18) is moral? For one thing, because I think it is; for another, because the fact that (18) is moral makes (19) moral, which in turn reveals the possibility of rejecting a descriptive metaphysical statement, (15), after rejecting a moral conclusion like (19). And this, it will be recalled, I indicated at the very beginning of this section. However, it is worth adding that some thinkers might be more inclined to surrender a moral statement than one that they would characterize as analytic or necessary. I myself have no strong inclination in that direction since I am quite prepared to amend or surrender *so-called* analytic statements. Consequently, in showing that a statement of the connection between being necessitated and being morally judgeable is itself moral, I make it easier for *others* to surrender statement (18) when they reject (19) in response to their recalcitrant emotions. After rejecting (19), I am quite prepared to save (18) and make alterations elsewhere in the reasoning from (15) through (19).

Once we view various combinations of metaphysical, moral, and logical views as alterable conjunctions that are to be tested by evaluating their capacity to organize sensory experiences and emotions, we cease to think of any one of the combinations as *the* combination that we *must* accept. And adopting the view that Kant's dictum about "ought" and "can" is moral should make this step in the direction of flexibility much easier for some. This shows that the alleged incompatibility between determinism and the practice of passing "ought"-judgments is more easily denied than might be supposed by those who regard the incompatibility as logical and who also think that logical principles are undeniable and beyond alteration or surrender.

In the light of the foregoing, one can imagine different philosophers trying to link their experiences and emotions by means of different mixed bodies of belief. In order to describe some of these bodies quickly and perspicuously, let us call determinism "D", let us use the letter "O" to denote the principle that "ought" implies "can", and let us use *"M"* to denote the principle that moral "ought"-judgments may be made. In that case, one philosopher might assert D and O but deny $M;$ a second philosopher might deny D while holding on to O and $M;$ and a third philosopher might assert $D,$ deny $O,$ and assert M. All of this assumes that no one of these philosophers tampers with a certain logical principle or tries to reinterpret "necessitate" in any context. Under the circumstances, if these philosophers have a common fund of experiences and moral emotions which they are trying to organize, they will try to see which combination best organizes those experiences and emotions. The issue will then be settled by them in a corporatistic manner.

Consider first the conjunction of $D, O,$ and the denial of M. It is held by a philosopher who accepts determinism, the moral dictum that "ought" implies "can", and the doctrine that we ought not make moral judgments. His determinism tells him that every event is necessitated and his moral dictum tells him that no necessitated action (which is to say *no* action if every action is an event) ought to be judged morally. Therefore, he is morally obligated to refrain from making moral judgments. Compare him with the philosopher who denies D while asserting O and M. This philosopher asserts that we *may* make moral judgments and that whenever we do make one, the judged action is *not* necessitated. That is why he surrenders determinism. Finally, consider the philosopher who accepts D and M but surrenders O. He thinks that every event is necessitated and that we may make moral judgments but, in order to avoid an inconsistent triad, he gives up the dictum that making a moral judgment about an action morally requires the action not to be necessitated. On the assumption that we do not question any standard logical principle, each one of these triads is internally consistent but they are all incompatible with each other. How should we decide which of them to accept? By seeing which of them most effectively organizes our experiences and moral emotions. This comparison might require us to appeal to the relative simplicity of the different triads and also to the degree to which they require us to abandon beliefs that we hold very firmly. In other words, all of the triads might link our experience and emotions but one might do so more effectively on the score of simplicity or of a desire to hold on to familiar principles.

12. *Freedom of Speech and Freedom of Belief*

Throughout the foregoing argument I have made the assumption that we are free to adopt, alter, or abandon our beliefs that certain statements are true. Such an assumption may become clearer if we recall a few episodes in the history of philosophy.

In his (first) *Letter Concerning Toleration,* Locke wrote: ". . . to believe this or that to be true does not depend on our will". By this he meant that we can *never,* by choosing, believe that a statement is true, whereas sometimes we can, by choosing, raise our arms. Thomas Jefferson echoed this view in his "Bill for Establishing Religious Freedom" when he said that he was "well aware that the opinions and belief of men depend not on their own will, but follow involuntarily the evidence proposed to their minds". The view was that once evidence has impinged on the mind, willing plays *no* part in bringing about belief. Moreover, willing cannot prevent evidence from producing a belief in us.[3]

About two hundred years after Locke's (first) *Letter Concerning Toleration* had appeared, William James published his famous essay, "The Will To Believe". Partly under the influence of Charles Renouvier,[4] James challenged Locke's view, but it is important to realize that James did not abandon it completely. In "The Will To Believe", James admits that we cannot, just by willing it, believe that Abraham Lincoln's existence is a myth; that we cannot, by an effort of will, believe ourselves to be well when we are "roaring with rheumatism in bed"; that we cannot successfully choose to believe that the sum of the two one-dollar bills in our pockets is a hundred dollars. He adds that Hume was correct in holding that some of our beliefs are true because of certain matters of fact whereas others are true because of certain relations of ideas. These matters of fact and these relations of ideas "are either there or not there for us", and if they are not there, they "cannot be put there by any action of our own". By contrast, James says, there are some statements which we are not

3. See Locke, *Works* (reprinted Darmstadt, 1963), Volume VI, pp. 39-40; *Papers of Thomas Jefferson,* ed. J. Boyd (Princeton, 1950), Volume 2, p. 545; Morton White, *The Philosophy of the American Revolution* (New York, 1978), pp. 198ff.
4. R. B. Perry, *Annotated Bibliography of the Writings of William James* (New York, 1920), p. 33.

forced to accept by matters of fact or relations of ideas, and consequently our will may, indeed must, play a part in accepting them as true.[5] In "The Will To Believe", therefore, James was not prepared to say that we can choose to accept any statement whatever as true. He limits the power of the will in this regard and holds that we may exercise our "will to believe" only in some cases. James replaces Locke's view that we can *never* choose "to believe this or that to be true" by the view that sometimes we can and sometimes we cannot, but at a later point I think that the will is given even greater scope by James himself.[6]

That point arrives when the scientist is viewed as one who can *always* assign truth-values to statements as he will, without being *forced* by Humeian matters of fact or relations among ideas to assign these truth-values in a certain way. According to this view, the scientist is epistemically obliged only to assign truth-values in a way that will link experiences successfully; he is not viewed as one whose mind contains picture-beliefs that are caused to mirror facts "as they really are". We may get some further understanding of how this view arose by noting that although James denies that we can, by an effort of will, *believe* in the truth of the statement that we are well when we are in bed with rheumatism, he admits that we can *say* that it is true that we are well. Here James gives expression to the idea that *saying* that something is true is importantly different from believing that something is true insofar as the former is external whereas the latter is internal. Therefore, saying that something is true is like kicking and like raising one's arm in an important respect whereas believing that something is true is not. This is related to Locke's idea that religious worship, which is external, can be performed through an act of will, whereas religious belief, an operation of the understanding, cannot be because it involuntarily follows the evidence. James held that *saying*, like worship, is external by Locke's standards and therefore that we can successfully will the *saying* that "Two dollar-bills add up to two hundred dollars" is true even though we cannot successfully will the *believing* that two dollar-bills add up to two hundred dollars.

5. See William James, *The Will To Believe and Other Essays in Popular Philosophy* (New York, 1897), pp. 4-5.
6. See my *Science and Sentiment in America: Philosophical Thought from Jonathan Edwards to John Dewey* (New York, 1972), pp. 204-16.

However, once we view codified science as a conjunction of *external sayings* that certain sentences are true and not as a conjunction of *internal believings,* the will is given far greater scope than Locke gave it or than James gave it in "The Will To Believe". Once science is viewed as a tool or man-made fabric, all of science must be viewed as a product of the will. Men have made the fabric because they have chosen to make it, and they can alter it as they choose.[7] True, they observe certain rules in making and altering the fabric, but they freely observe them. Even the decision to construct one that accommodates experience is a free decision.

I hope that this helps the reader to understand my position more clearly. I have been working in the tradition of what may be called voluntarism since I believe that the hypothetico-deductive arguments I have been talking about are freely built and freely alterable linguistic structures which serve as tools for linking sensory experiences with each other in some cases, and for linking sensory experiences with moral feelings in others. We are not *forced* by experience or feeling to build any one of these structures just as we are not forced by clay to fashion it into a pot. This is the assumption I have tried to state and clarify in the above historical digression. I hope it will serve to illuminate what I have been saying so far about mixed bodies of statements, and especially about our right to alter them in certain ways when they contain moral normative statements about actions. I also hope that it will prepare the way for what I am about to say about mixed bodies of statements that contain what I call epistemic or epistemological normative statements. Since *saying* is an act we can successfully choose to perform, we may make epistemic normative statements about acts of saying.

This is a good place at which to repeat that so long as one regards the problem of what one may or ought to believe as a problem of what linguistic expressions one may or ought to accept in one's effort to organize experience and emotion, one

---

7. Within the last generation the view of science as a man-made fabric has been emphasized by Quine, *From a Logical Point of View,* p. 42, and also by Goodman, *Ways of Worldmaking,* esp. Chapter 6. James advocated much the same view at the turn of the twentieth century; see my *Science and Sentiment in America,* pp. 338-39, nn. 61 and 75.

avoids the view that there are ready-made facts which a rational believer is obliged to mirror or copy. For once we assert the existence of such facts, we veer away from corporatism. If we assert the existence of facts which a believer is obliged to mirror or copy before attributing truth to a statement, we are committed to a statement-by-statement check of individual statements and cannot accept corporatism. Each individual true statement would have to mirror a fact, in which case it would not be possible to exchange one conjunction of statements for another without seeing whether each conjunct mirrored a fact. We would have to look at these facts one by one in order to establish statements one by one. By contrast, I think we may dispense with such "facts" in assessing our linguistic structure as a tool for linking experience and emotion, that is to say, dispense with asserting the existence of extra-linguistic rails in nature to which individual statements must stick.

It might be argued in response that if we dispense with facts, we dispense with data; and that if there are no data, then there is nothing to be organized or linked by science or ethics. However, this reaction may arise from an ambiguity of philosophical language. A fact may be regarded as the denotation of a statement like "Mandy is on the mat" or as the denotation of some proper part of that statement such as its subject, "Mandy". If a fact is regarded in the latter way, then I have no objections to facts. However, if a fact is regarded as the denotation of a statement like "Mandy is on the mat", then the fact that Mandy is on the mat is an abstract entity which is quite unlike Mandy the cat and quite mysterious. You can lift Mandy but you cannot lift the fact that Mandy is on the mat. Consequently, when I said that I dispensed with facts, I was merely trying to say that I dispensed with references to abstract entities like the fact that Mandy is on the mat. This view, however, is compatible with accepting the existence of Mandy, and of sensory experiences and feelings which form the data of science and ethics. *They* may be called "facts" along with Mandy but they are not abstract entities and they may be linked or organized. For this reason there is no inconsistency in holding that science and ethics organize sensory experiences and feelings while dispensing with the assumption that there are abstract entities to which *statements* refer or correspond.

13. *Normative Statements in a Metalanguage:*
Moral and Epistemic

The normative conclusion (19)—"No human action ought to be judged morally"—as well as the normative premise (18)—"If a human action is necessitated, it ought not to be judged morally"— both speak about *judging* actions morally, but the point about the argument in which these statements appear may also be made by using as an illustration an argument in which statements about so-called "speech-acts" appear. In other words, statement (18) may be replaced by the following:

> (18') If a human action is necessitated, we ought not to say "It ought to be done", "It ought not to be done", or "It may be done".

And, of course, (19) may be similarly replaced by

> (19') Of no human action ought we to say "It ought to be done", "It ought not to be done", or "It may be done".

In that case our illustrative moral argument is transformed into one whose conclusion forbids certain speech-acts rather than certain judgments or beliefs. Moreover, statements (18) and (19) are transformed into constituents of a metalanguage: they become statements about statements.

Because they are *moral* metalinguistic statements, along with the statement that no one ought to lie, it may be asked whether *all* normative metalinguistic statements are ethical in character. For example, is the corollary of the corporatism that I have advocated, namely, "We may reject a descriptive premise in a mixed body of premises in response to our rejection of a normative conclusion", an ethical statement? Or, turning to statements that sometimes appear in works of formal logic, what about the statement in which we forbid the affirming of the consequent: "We may not infer 'Socrates is a man' from 'All men are mortal' and 'Socrates is a mortal' "? Is *it* an ethical statement?

Let me begin by saying that I am not able to provide a clear definition of the phrase "ethical statement" which will help us distinguish ethical metalinguistic statements from other metalinguistic statements in which expressions like "ought", "ought not", and "may" appear. Indeed, I think that the effort to define the name of a discipline is often unsuccessful or trivial in its outcome. Nevertheless, I am sure that most philosophers would draw a line between the metalinguistic normative principle that forbids the deliberate telling of a falsehood and the metalinguistic normative principle that forbids affirming the consequent. But what is the relevant difference between these two prohibitions? I am not able to give an answer that satisfies me completely but I venture to say that the fact that a lie is deliberately told with the intention to deceive is what makes it the concern of the moralist. By contrast, a fallacious affirmation of the consequent may be a slip and therefore not of concern to the moralist. An affirmation of the consequent is not *as such* fair game for the moralist whereas a lie is—whether the moralist should view it favorably or adversely. This is not to say that some sophistical reasoners have not argued fallaciously with the intention to hoodwink someone, but we can argue fallaciously without the intention of hoodwinking someone. We may do it inadvertently and we may therefore regard the injunction against *such an argument* as non-moral.

Having maintained that some normative metalinguistic statements are ethical and that some are not, I now want to say that the metalinguistic thesis of the earlier part of this inquiry—"We may reject a descriptive premise in a mixed body of premises after having rejected its normative conclusion"—is *not* ethical. Therefore, when we say "We have a right to reject the statement 'The mother killed a human being' after having reached the conclusion 'The mother had a right to do what she did' ", the statement of a right within double quotes is *not* ethical whereas the statement of a right within single quotes *is* ethical. And if we had formulated our thesis by speaking about beliefs rather than about statements, we should say that the belief that the mother had a right to do what she did is an *ethical* normative belief whereas the belief that we have a right to reject a descriptive belief in the wake of rejecting a normative belief is an *epistemic* normative belief. Furthermore, I should say that it is an epistemic normative belief that is not equivalent to an ethical belief. The net effect of this is to put me in the following position with regard to the bearing of normative ethical beliefs on our descriptive beliefs. I have said that an ethical normative statement may be permitted to affect what descriptive statements we make in an argument like that about abortion but I do not regard the very statement in which I say this as an ethical normative statement. I do not think that my epistemic statement is in a crucial respect like the statement 'A man may kill in self-defense".

Since I have distinguished between normative statements that are ethical and those that are not, it might be thought that I am forced—in spite of my doubts about the notion of meaning—to hold that the words "ought" and "may" mean different things in ethical statements from what they mean in epistemological statements. Thus it might be held that I am forced to say that "One may lie to save the lives of those one loves" contains an occurrence of the word "may" that is not synonymous with the same word as it appears in the statement "You may reject a descriptive premise in a mixed body of premises after having rejected its normative conclusion". But I am not forced to hold this view, just as I am not forced to hold that the word "exists" means one thing in the context "Physical objects exist" and another in the context "Universals exist".[8] And, analogously, when one uses the word "may" in statements which assert that one may lie under certain circumstances, one is not forced to hold that the *meaning* of "may" is different from its meaning when used in metalinguistic statements which assert that one may, under certain circumstances, reject a descriptive statement after one has rejected a moral normative statement. Nevertheless, when we say that someone ought to perform a certain action of the kind that is judged in ethics, we may say that the action is *morally obligatory;* and when we say that a man has an obligation to accept a certain conjunction of statements, we may say that it is *epistemically obligatory* that he accept it. If we speak in this way, we make clear that there is a genus of obligation that contains at least two species. We may also distinguish morally permissible and epistemically permissible acts, thereby indicating that there is a genus of permissibility that contains at least two species.

If one proceeds in this way, one may distinguish an ethical statement containing the word "ought" or the word "may" by reference to *the kind of action* that is said to be obligatory or permissible. If the action is classified as one of lying, or stealing, or others mentioned in the Ten Commandments, for exam-

8. See my *Toward Reunion in Philosophy,* Part I.

ple, then the normative statement is ethical. And if the action is not so classified, the normative statement will not be ethical. The difficult question is whether one can go any further than the characterization "like lying, or stealing, or others mentioned in the Ten Commandments", and I am not sure that one can go much further. Nevertheless, I do think it important, as I have already indicated, to maintain the distinction between an ethical normative metalinguistic statement and a non-ethical one if only to prevent a thesis like that of corporatism from being *automatically* viewed as ethical just because it contains a word like "may" or a word like "ought". The fact that the genus to which the predicate "obligatory" refers contains the class of obligatory beliefs whose obligatoriness is asserted in epistemology as well as the class of obligatory actions whose obligatoriness is asserted in ethics shows that the predicate "obligatory" is neutral as between epistemology and ethics. That is why the statement that we ought not to make a certain statement is not necessarily an ethical statement. It may be when we use it to proscribe a mendacious statement, but not when we use it to tell a physicist that the evidence requires him not to make a certain statement. Therefore, "You should not have made that statement because in making it you lied" is importantly different from "You should not have made that statement because in making it you said something for which you had insufficient evidence". Yet the difference between these two "should"-statements does not depend on any difference in the meaning or sense of "should"; it depends on the different reasons that are given for forbidding the making of the very same statement.

What I have been saying reveals a difference between my view and that of William James in his *Pragmatism*. He says there "that truth is *one species of good,* and not, as is usually supposed, a category distinct from good and coordinate with it",[9] but when he speaks of *good* he does not seem to have in mind a generic notion which is neutral as between ethics and epistemology. In denying that the true and the good are coordinate categories and in making truth one species of good, James seems to treat the good as ethical. On the other hand, the traditional doctrine he opposed, namely, that the true and the good are coordinate, has

---

9. William James, *Pragmatism* (reprinted Cambridge, Mass., 1975), p. 42.

it that the true is not ethical.[10] But James tells us that "what would be better for us to believe" sounds very much like a definition of truth which comes very near to defining the truth as "what we *ought* to believe", the latter being a definition in which his readers, he declares, would not find any oddity.[11] And I think that James would have regarded the "ought" in this definition as ethical. In any case, whether or not I am correct in this interpretation of James, I want to make clear that I do not think of epistemology as a chapter of ethics.

Why? Because ethical language is used in order to organize experiences and moral emotions of a sort that the normative epistemologist does not try to organize. In saying this, I anticipate a thesis that I shall defend later, namely, that the normative epistemologist *does* try to link emotions and experiences. Therefore, the most I want to say here is that the emotions or feelings which we have when we contemplate the kinds of actions judged by normative moralists are, generally speaking, different from those which we have when we contemplate the kinds of actions judged by epistemologists. In other words, the feeling of moral obligation or of moral entitlement is different from the feeling of epistemic obligation or of epistemic entitlement. My feeling obligated not to tell a lie or to steal is different from my feeling obligated not to adopt the belief that snow is not white. How these remarks affect other things concerning epistemic duties and rights will become evident later on.

The reader will have noticed that I have distinguished between normative moral discourse and normative epistemic discourse by distinguishing (1) the different *sorts of actions* that are judged by the moralist and the epistemologist; (2) the different *reasons* that are given by them for commanding, permitting, or forbidding actions; and (3) the different sorts of *feelings* that the moralist and the epistemologist have when contemplating the actions they judge. Let me try to show how all of these distinctions are related.

Distinctions (1) and (2) are very closely related. When we say

10. See, for example, Victor Cousin's remark, "The idea of the good is the whole of ethics", *Lectures on the True, the Beautiful, and the Good,* trans. O. W. Wight (New York, 1854), p. 215.
11. James, *Pragmatism,* p. 42.

that a certain action ought not to be done because it is an act of stealing, our *reason* is that it is an *act of stealing*. By contrast, when we say that a certain inference ought not to be made, our *reason* may be that it is an *act in which we affirm the consequent*. In each case our reason involves describing the prohibited act in a certain way, but our descriptions differ in kind just to the extent to which stealing differs in kind from affirming the consequent. (Note, however, that an affirmation of the consequent may as such be condemned logically and, in addition, condemned morally if it has been performed with the intention to deceive.) And when we compare all the acts that resemble stealing in a crucial respect with all the acts that resemble affirming the consequent in that respect, we shall, I hope, be able to see what difference in kind there is between moral acts and epistemic acts. Finally, about distinction (3) I have merely tried to say that the moralist's feeling of obligation not to steal is quite different from the epistemologist's (or the logician's) feeling of obligation not to affirm the consequent. They are both negative feelings, but I may give some idea of how I conceive their difference by saying that they are as different from each other as a feeling of rage and a feeling of discomfort upon hearing a jarring noise.

14. *Limited Corporatism in the Metalanguage*

So far I have concentrated on the implications of limited corporatism for mixed arguments that contain normative ethical statements as well as non-normative statements that describe actions of the kind with which moralists are concerned. I have sometimes spoken of such arguments as made up of statements and at other times I have spoken of them as made up of the beliefs that are expressed by these statements. So far it has not been necessary or useful to settle on one of these ways of speaking but it will be convenient to confine myself for a while to speaking about statements in trying to compare and contrast the role of normative thinking in ethics and epistemology.

Even though I have said that a normative metalinguistic principle such as "We may amend or surrender a descriptive premise in a mixed set of premises after rejecting a normative moral conclusion" is not moral, it shares something important with normative moral principles in addition to being normative. Like an ethical normative principle, it is not to be tested in isolation but rather as part of a conjunction of statements, some of them normative and others descriptive. It follows that the relationship between the normative epistemology of inquiry and the description of inquiry is in certain ways similar to that between the ethics of action and the description of action. Or, to state the point in yet another way, the connection between the theory of knowledge or science and the history of knowledge or science is in a certain respect similar to that between the ethics of action and the description of action. I mean that while thinking on a higher linguistic level *about* descriptive science or *about* moral reasoning, we engage in reasoning which is similar to moral reasoning itself. For in speaking or writing about certain features of a practical *ethical* argument, we use normative principles about what we ought to do, ought not to do, or may do in altering our first-level arguments under certain circumstances; we describe what a certain individual has done; and we deduce logically singular conclusions from a conjunction of logical principles, normative principles, and descriptive statements of history or psychology. I emphasize that we do this while speaking or writing about first-level ethical arguments which are mixed *and* about first-level arguments that are made up exclusively of descriptive statements. Once we are operating on this higher level, we are operating in the metalanguage that is one linguistic step above that of ordinary ethical argument and ordinary physical argument. But I wish to emphasize that the normative principles of corporatistic epistemology on this level are not tested as individual statements. They are tested as components of conjunctions of second-level statements. The testing of second-level epistemology is just as corporatistic as the testing of the ethics or the physics it is about.

As an example that may clarify what I have in mind, let us consider Einstein's view that physical theory should be deterministic. This may be regarded as a metalinguistic statement to the effect that only a deterministic physical system ought to be accepted as a really adequate explanatory system. Upon examining contemporary physical theory and saying that it is not deterministic, it is obvious that a follower of Einstein, reasoning as we did earlier in our ethical examples, would deduce that contemporary physics ought not to be accepted as final. He would argue as follows: "No non-deterministic physics ought to be accepted as final; contemporary physics is non-deterministic; therefore, contemporary physics ought not to be accepted as final". Now let us imagine that a contemporary physicist rejects the conclusion for reasons that we need not discuss at the moment. Obviously, the contemporary physicist will be faced with a question the form of which has often been discussed in earlier pages: Should he deny the major premise or should he deny the minor descriptive premise? He might give up Einstein's normative principle that non-deterministic physics not be accepted as final and hold on to the descriptive statement that contemporary physics is non-deterministic, or he might do the reverse if he is not moved to surrender both premises. As in the case of the ethical examples considered earlier, the major premise of the epistemological argument cannot be certified on the basis of the usual array of attributes that are cited by those who wish to confer absolute certainty on principles: self-evidence, analyticity, a prioriness, and other attributes that are associated with being immune to revision or rejection. Therefore, the physicist who denies Einstein's singular normative conclusion is not bound to accept Einstein's major premise, just as the moral judge is not bound to treat moral principles as irrevocable. I want to underscore the fact that this same physicist may reject the descriptive premise that contemporary physics is not deterministic. Such a move would be comparable to denying in our moral illustration that every fetus is a human being; it would constitute an effort to save Einstein's major premise after denying the conclusion of his argument. There are those who would say that such a move might be made only by surrendering an allegedly analytic statement about what determinism is. But clearly the corporatist would have no qualms about this, being prepared to surrender even allegedly analytic statements when he tests the conjunctions of which they are components.

So far, then, we have risen to the metalanguage of physics and presented a mixed argument about a physical system. And once we set forth this second-level argument, we find that it resembles a mixed first-level argument. Returning now to the Einsteinian epistemic argument, I want to point out that it is more common to reject or amend its major normative premise than it is to reject or amend the minor descriptive premise which asserts that contemporary physics is not deterministic. And this practice may reflect the acceptance of a maxim which says that normally when one is faced with an argument having the form of Einstein's epistemic argument and one rejects its conclusion, one should reject its major premise rather than its minor. After all, its major premise is a methodological rule which is similar to many others that have been jettisoned in the history of physics, for example, the rule that we should accept only mechanistic theories. It appears, therefore, that when we confront analogous choices in the case of the moral practical argument and the epistemic practical argument, we are not guided by similar maxims while trying to decide what to give up and what to save in the event of a crisis. For it does seem that a previously accepted rule of morality has a standing by comparison to any so-called statement of fact which would predispose us to give the former up only under the most extreme circumstances when there is a contest between giving it up and giving up the latter. By contrast, when there is a similar contest with regard to an epistemic argument, the rules of scientific method are not viewed with the same degree of respect—especially by scientists.

Rules of epistemology or methodology have often been swept aside by working scientists and rules of morality have been swept aside by ordinary moral judges, but previously accepted rules of morality may be more respected by ordinary moral judges than previously accepted rules of scientific method are respected by scientists. Certain principles of the Decalogue have not survived the ravages of social change but I doubt that ordinary men have rejected any principle of morality more decisively than twentieth-century physicists have rejected the rule requiring them to be determinists. Nevertheless, whatever the contrasts between the maxims that may govern our decisions to alter the premises of moral practical arguments and those that govern our decisions to alter the premises of epistemic practical arguments, we are dealing in each case with a mixed body of statements which leads to a normative conclusion that may be rejected; and such a rejection requires us to decide whether we should drop a normative rule or a descriptive statement. Limited corporatism applies to epistemic as well as moral practical arguments.

15. *Testing the Premises of Epistemic Normative Arguments*

The following objection might be raised against my view that we test practical epistemic statements in a corporatistic manner. According to my view, we rise at this point to a metalanguage in which we advance mixed epistemic arguments whose premises are *about* first-level statements of descriptive science or *about* first-level statements of moralists. Consequently, it might be argued that we cannot test such a mixed body of premises concerning statements by assessing its capacity to link sensory experiences and moral feelings like those I mentioned when I was discussing moral practical arguments. There, it might be said, it was plausible to speak about our emotional reactions to acts of lying and equally plausible to speak of the sensory impressions we have of such acts. But what happens when we begin to speak about speech-acts that are not of concern to moralists? Once *they* become the subject-matter of our metalinguistic normative rules and conclusions or the subject-matter of our metalinguistic descriptive statements, the situation seems to change. Since we no longer react to human actions that are said to be *morally* obligatory or permissible, we may find it difficult to think that we are linking sensory experiences and the feeling of obligation. In a moral practical argument we speak about human actions that may be described and also called obligatory by appealing to our experiences and moral feelings, but how, it may be asked, can we respond in this way when speaking about speech-acts that are of no moral interest?

To be clear about the issue, we may think of a body of physical theory as a body of speech-acts. The comparison becomes even more perspicuous if we think of a body of physical theory as one act of accepting a long conjunction. In that case, if we compare a moral normative principle like "No act of taking a person's life ought to be performed" with an epistemic normative principle like "No body of physical theory that is not deterministic ought to be accepted", we may say that the subject-matter of the former, "act of taking a person's life", is different in an important respect from its counterpart, "body of physical theory", and that the predicate-term of the former, "ought to be performed", is different in an important respect from "ought to be accepted". Nevertheless, the mentioned moral normative principle and the mentioned epistemic normative principle both contain descriptive subject-terms and they both contain normative predicate-terms. That is why, when testing corporatistically in epistemology, we must try to see whether a mixed conjunction of metalinguistic premises links our sensory experiences and certain feelings. For there is as much of an appeal to sensory experience in describing our long physical speech-act as non-deterministic as there is in describing an act as one of taking a person's life: one must observe our speech-act to see whether it is non-deterministic. Moreover, there is as much of an appeal to feeling associated with asserting that a body of theory ought or ought not to be accepted as there is in asserting that an act of taking a person's life ought or ought not to be performed. Therefore, when we come to decide whether or not we ought to accept a mixed second-level conjunction, we must refer to sensory experiences and feelings about *speech-acts* even though we are no longer directly concerned with acts that are *morally* judgeable, such as acts of stealing. For this reason, even though the premises of an epistemic practical argument are about first-level linguistic acts, we test those metalinguistic premises by methods that are analogous to those that we use when testing the premises of moral practical arguments that are about acts of stealing. We see whether a conjunction of metalinguistic premises successfully links experiences and feelings, but the experiences and feelings are different from what they are in ethics.

The view I advocate may bring to mind Hume's view that belief is a feeling or sentiment.[12] But I go beyond that and maintain that there is a *feeling of being obliged* to accept a statement and a *feeling of having a right* to accept a statement. I also hold that these feelings, which we take into account when we decide to accept or reject a second-level conjunction about first-level descriptive statements, may in turn be linked with experiences that are germane to certain statements about first-level statements. For example, those physicists who disagree with Einstein and who accept a non-deterministic physics feel a duty or at least a right to accept it that does not come out of the blue. Their feelings will be connected with certain experiences of theirs concerning non-deterministic physics. Presumably the anti-Einsteinians will

12. *A Treatise of Human Nature,* Appendix, p. 624.

have tested the body of non-deterministic theory in a corporatistic manner and will have concluded that because it has certain descriptive features, they feel a duty or at least a right to accept the theory. That feeling they express when they reject the normative conclusion deduced by Einstein from the premises of his epistemic practical argument. But even though the anti-Einsteinians' feelings and their experiences concerning non-deterministic physics are different from the experiences and feelings that we have concerning morally judgeable acts, this does not gainsay the fact that physicists have feelings of obligation to accept statements just as moralists feel obliged to do things that form the subject-matter of a moral practical argument.

In order to amplify what I have just been saying I shall suppose for the sake of illustration that we have accepted the premises of a moral practical argument because the conjunction of them effectively links what we may call certain first-level experiences and feelings about a moral action. For example, we may imagine that we have accepted the conjunction of premises of our syllogism composed of (6), (7), and (8) because certain experiences and moral emotions have been linked or organized by it. When we defend our acceptance of the conjunction of (6), (7), and the relevant logical law, we will say that any first-level conjunction which accomplishes such linking ought to be accepted and that this first-level conjunction does accomplish this. However, the principle we use in this defense is a normative principle of epistemology, and when we defend this second-level principle we will, in keeping with corporatism, defend it as part of a second-level conjunction that links second-level sensory experiences and feelings that we have when contemplating beliefs or statements.

These observations show that epistemological normative principles govern the acceptance of bodies of statements much as moral practical principles govern the performance of actions that are assessed by moralists. Furthermore, epistemological normative principles are not immune to revision, thereby displaying another feature that they share with moral normative principles. Like moral principles, they may be amended or surrendered in the light of what may be inferred from the conjunction of them and certain metalinguistic descriptive statements. And also like moral principles, they cannot be supported by appealing to dubious notions like analyticity, logical necessity, a prioriness, or self-evidence. I do not think they can be defended by appealing to the definition or essence of science or to that of some particular science, any more than I think that moral principles can be defended, in the manner of the doctrine of natural law, by appealing to the definition or essence of man. If one were to appeal to the essence or analysis of being a scientist, one might say that to be a scientist is to be an individual who accepts or rejects bodies of belief in accordance with specifically mentioned epistemic normative principles. In that case a violation of these principles would lead one to say that the violator was *necessarily* not a scientist just as certain advocates of the doctrine of natural law— for example, Locke—hold that a person who violates certain moral principles is necessarily not human because that person has degenerated from his status as a man. But I cannot accept Locke's view and I cannot accept the analogous view that appeals to the essence of a scientist.[13]

The latter view would regard a scientist as one who, by virtue of the essence of being a scientist, operates in accordance with specifically mentioned principles of scientific method. Yet these principles might well be discarded with the passage of time and the development of science.[14] Therefore, if a scientist is conceived as one whose work is essentially governed by the principle that all science ought to be deterministic, physicists who today dispense with determinism would have to be regarded as nonscientists. It might be said in reply that although it is a mistake to identify the essence of being a scientist by reference to the acceptance of so narrow a principle as that of determinism, other identifications might not prove as restrictive. In this spirit, someone might try to combine essentialism and corporatism by saying that a descriptive scientist is essentially one who tries to create intellectual devices that link our first-level sensory experiences and that a moralist is essentially one who does this and also tries to link our first-level experiences and emotions concerning moral actions. What about these views? They seem very modest and tempting. They do not tie the essence of science to anything as specific as determinism but merely to the most general features of science as I myself have depicted it. Can I consistently take

13. See my *Philosophy of the American Revolution,* pp. 170ff.
14. For an illuminating discussion of this topic see Dudley Shapere, "The Character of Scientific Change", ed. T. Nickles (Dordrecht, 1980), pp. 61-116.

exception to them? Yes, so long as the notion of essence (or that of meaning) is used in formulating them. If we say that a descriptive scientist is one who *essentially*, or by virtue of the meaning of the term "scientist", ought to create devices that link sensory experiences, we use the objectionable notion of essence or the objectionable notion of meaning in bestowing or denying the title "scientist". On the other hand, since I think that descriptive science should corporatistically link sensory experiences and that morals should corporatistically link sensory experiences and emotion, how can I be consistent? By formulating my view without using the concept of essence or allied concepts.

Although I refrain from using the notions of meaning and essence, I think I understand and sympathize with the intentions of some philosophers who employ these notions. Keeping these intentions in mind, I am quite prepared to grant that we are properly more reluctant to surrender some statements that figure in certain conjunctions than we are to surrender others, and that what are called analytic statements or essential truths coincide to some extent with those statements that we have a right or duty to protect to the bitter end in any inquiry. I think this true not only of certain statements in descriptive science but also of some in epistemology, which in my opinion contains at least two principles of this kind: the corporatistic principle that every descriptive scientist ought to link sensory experiences with each other and the corporatistic principle that every normative scientist ought to link certain sensory experiences with the feeling of obligation. But, by refusing to regard even these principles as analytic or as essential truths, I avoid viewing the rules that govern scientific behavior as immutable just as I avoid viewing the rules of morality as immutable. I leave open the possibility that future descriptive scientists, the professional descendants of the practitioners of the discipline that we call descriptive science today, might change their practice so as to require the abandonment of even the most revered principles of methodology or epistemology. In short, I hold that the principles of normative epistemology are no more sacrosanct than the principles of descriptive science or the principles of normative ethics.

After having said this, I might well be asked to describe a situation in which I would abandon corporatism. In other words, I might be asked to describe a situation in which I would not test

a statement by conjoining it with other statements and then seeing whether the resulting conjunction successfully linked sensory
experiences or sensory experiences with certain feelings. In reply,
I should say that I find it hard to describe such a situation by the
use of terms that I regard as clear, but I do not think that my
failure to provide a comprehensible description of a situation in
which I would surrender corporatism shows that corporatism is
not surrenderable. The statement "Corporatism is surrenderable"
does not imply the statement "We are able to state with clarity
the conditions under which corporatism would be surrendered".
Therefore, "We are *not* able to state with clarity the conditions
under which corporatism would be surrendered" does not imply
"Corporatism is *not* surrenderable". For these reasons I feel no
great qualms about saying that corporatism itself is surrenderable
even though I am unable clearly to specify conditions under
which it would be surrendered. On the other hand, if the philosopher who asks me to specify such conditions is trying to see
whether there is a philosophical theory—however obscure—that is
opposed to corporatism, the answer is obvious. The chief oppo-
nent of corporatism is the dubious doctrine that we may or ought
to test each of our statements, whether in descriptive or normative science, individually and in isolation; that we ought to con-
sult different realms of being in order to test different kinds of
statements; that we use different faculties in testing them; that
"truth" means different things when applied to different kinds
of statements, and so on through the history of intellectual frag-
mentation that has flourished for two millenia of Western philosophy. In short, one argument for the surrenderability of corporatism is the fact that arguments for its not being surrenderable
depend on the use of excessively obscure statements about how
statements ought to be tested.

V

The Good in Ethics and Epistemology

16. The Role of the Good in Ethics

So far I have concentrated on normative statements in epistemology and ethics; and because I stipulate that a normative statement is one that expresses a duty or a right to do something, I have said very little about other statements to which other philosophers have paid much attention, statements in which we say that certain states of being are *good*. Many philosophers hold that there is a very important connection between such statements and what I call moral normative statements. For example, some philosophers hold that every normative statement about a duty to perform a certain action is synonymous with a statement that the action will produce the greatest possible amount of good in the universe, but I cannot accept this view because of my doubts about synonymy. I also decline to accept the view that when we say a moral principle is true or acceptable we *mean* that the general practice to which it refers maximizes the amount of good in the universe. Therefore, I may be fairly asked whether I believe that the notion of good plays any part at all in normative thinking.

Let me begin with a typical case in which the notion of good does *not* play a part in the process of supporting or rejecting a singular normative conclusion. In what I shall call the standard case of moral reasoning, a moral conclusion like (5) is deduced from a standard moral principle in a code and an appropriate descriptive premise. Here the process of testing will not include

any reference to good consequences. Here the singular moral conclusion will be tested by conjoining it with a statement that the
agent is normal (and the concept of normality, as I have indicated
earlier, may vary), that he contemplated performing the action
while in normal circumstances, and that whenever a normal person in normal circumstances contemplates performing an act like
that mentioned in (5), that person will feel an obligation not to
perform the act. If, then, the sort of person mentioned does have
such a feeling, the singular moral conclusion will be confirmed
and this will give corporatistic support to the premises of an
argument like the one running from (1) through (5). And if the
agent does not have such a feeling, the premises will have suffered
a setback. Now, however, I want to consider what may be called
a non-standard argument, one in which a reference to good consequences may figure.

Let us suppose that a moral judge who wishes to defend an
action that he contemplates performing cannot defend it by appealing to an ordinary moral principle that expresses either a
right or a duty. In such a case a moral judge must complicate his
moral principle in very many ad hoc ways in order to be able to
give a reason for his action and to organize his feelings and experiences. Suppose, for example, that a tyrant is hospitalized and
dying of cancer in a country other than the one he has tyrannized
and that he cannot get the medical attention he needs anywhere
else. Suppose, too, that this tyrant has at one time aided the
country in which he is dying. The tyrant's native country, now
ruled by a second tyrant, demands that the first tyrant be returned to his native country so that he may be tried. Let us further suppose that the second tyrant threatens to kill a number
of citizens of the country in which the first tyrant is hospitalized
unless the first tyrant is returned. And, to complicate matters still
further, let us suppose that the second tyrant threatens to cut off
the supply of much-needed oil to the country in which the first
tyrant is hospitalized. Now let us suppose that the moral judge
wishes to defend a singular moral conclusion in this situation. I
need not specify which. Let it be either the conclusion that the
first tyrant ought to be returned or that he ought not to be returned. And let us imagine that the moral judge wishes to defend that conclusion by appealing to a standard moral principle

or to standard moral principles and that none is to be found that will do the job. Under such circumstances the moral judge may well appeal to the goodness of the consequences of the contemplated act.[1]

Such an appeal to the consequences of, say, returning the first tyrant—if that is the action which is said to be obligatory—may take the following form. A number of descriptive statements that are deemed to be relevant will be asserted. Each of these will say something about the first tyrant by describing his state, his actions, or his relationships. In addition there will be a general descriptive statement which asserts that whenever an individual has the features expressed in the above-mentioned singular statements, returning him to his native country for trial will produce better consequences than any other action that we could perform. From all of these purely descriptive statements we may deduce the statement that returning the first tyrant will produce more good than any other action that we could perform. But, having reached this singular *descriptive* conclusion expressing *goodness*— descriptive because it does not express a duty or a right—we must go further if we are to reach our singular *normative* conclusion expressing *obligation*. We must connect being an action which produces the most possible good with being an obligatory action by asserting something like this: "If an action does not fall under a standard moral principle but produces the most possible good in the circumstances, then it ought to be performed". Upon adding this premise we might be able to deduce the conclusion that

1. For a similar view see Bertrand Russell, "The Elements of Ethics", *Philosophical Essays* (London, 1910), p. 19; reprinted in W. Sellars and J. Hospers, *Readings in Ethical Theory*, in which see p. 11. I should add that although Russell here seems to hold that we may solve some moral problems by appealing to a standard, simple precept like "Thou shalt not steal" without referring to the goodness of the consequences of an action, he also holds that this *precept* may be justified only by consideration of consequences. Therefore, my views are only partly in agreement with his insofar as I do not hold that we test standard precepts by appealing to consequences in the manner advocated by Russell and by G. E. Moore, whose *Principia Ethica* had exerted great influence on Russell's thinking in "The Elements of Ethics". Although my corporatism requires us to derive the logical consequences of a conjunction of premises that contains moral precepts in an effort to see whether that conjunction links experience and moral feeling, this does not imply that we must test the *causal* consequences of actions in the manner of utilitarianism or what has also been called "consequentialism".

the first tyrant ought to be returned; and, of course, an appropriately devised set of alternative premises might suffice to imply that the first tyrant ought *not* to be returned.

The use of a premise linking the good and the obligatory has stimulated much discussion by moral philosophers who wonder about the status of such a principle. Some say it cannot be empirical because goodness and obligatoriness are not empirical concepts; others say it cannot be analytic because goodness is an unanalyzable concept which therefore cannot contain the concept of obligatoriness; still others are propelled by both of these views into asserting that it is a synthetic a priori truth.[2] I bypass all of these questions not only because of my doubts about the analytic and the synthetic but also because my version of the premise is not to be examined in isolation but rather as part of a body of statements which is to link experience and the feeling of obligation. In a case of the kind I have just discussed, the test comes when we examine the normative conclusion as we examine others. If, upon testing this conclusion corporatistically, we find that it is acceptable, then it passes its support back up the logical chain, so to speak, to the conjunction that implies it; and that conjunction contains the principle connecting the good and the obligatory. However, the testing of the normative singular conclusion involves an appeal to feelings of obligation, as it does in the case of standard moral arguments. The principle linking the good and the obligatory is corporatistically tested as a standard moral principle is tested, that is to say, as part of a conjunction of statements that links our experiences and the feeling of obligation to perform a certain act.

This raises the question whether the principle linking the good and the obligatory is itself a moral principle, albeit not a standard one. Some, of course, will deny that it is for the following reason. They will say that a standard moral principle links a descriptive term like "adultery" with the normative term "ought not to be committed" whereas the principle that links the term "action which does not fall under a standard moral principle but produces the most possible good in the circumstances" with the term "obligatory" links two non-descriptive terms. However, it is

2. For example, see W. K. Frankena, "Obligation and Value in the Ethics of G. E. Moore," *The Philosophy of G. E. Moore*, ed. P. A. Schilpp, pp. 93-110; also Moore's reply to Frankena in the same volume, pp. 554-81, 592-611.

not at all obvious to me that the long term containing "good" is a non-descriptive term or that singular sentences which contain the word "good" are non-descriptive sentences. I realize that there are philosophers who regard "good" as synonymous with "ought to be desired" and who would therefore assert that statements containing "good" are disguised "ought"-statements. But I question this reductionistic treatment of "good" just as I question reductionistic treatments of "ought". So the issue is whether we can distinguish usefully between a class of terms into which "good" falls and one into which "obligatory" falls without trying to rely on dubious "analyses" or alleged definitional equivalents of these terms.

From a certain point of view it does not really matter whether we call "obligatory" a normative term and "good" a non-normative term since the word "normative" is itself not all that clear when ordinarily used. Nevertheless, there is a difference between "good" and "obligatory" which may be conveniently marked by saying that the latter is a normative term whereas the former is not. The crucial point is that when we say that a certain action is obligatory (or permissible) we make the last guiding statement that we can make to a person (including ourself) who wishes to decide whether to perform the action. On the other hand, when we say to the same person that the act, though not falling under a standard moral rule, will produce good results, we have *not* made the last guiding statement we can make, since it is still open to him to ask whether he should (or may) perform the action. I do not deny that the advised person can reject both the statement that an action is obligatory and the statement that the, so to speak, unregulated action has good consequences. I mean only to say that after he has agreed that the unregulated action has good consequences, he may properly ask *another* question, namely, whether he ought to perform the action; whereas, after he has agreed that he ought to perform the action, he cannot properly ask this same question all over again unless he is wondering whether he was right to agree.

I now wish to return to the question that launched these terminological remarks. It is now clear why I cannot accept the view that whereas a standard moral principle links a descriptive term and a normative term, the principle that connects goodness and obligation links *two* normative terms. For me the term "good" is descriptive because it does not refer to what ought to be done, and therefore the alleged difference between standard moral principles and that which connects goodness and obligation disappears. The statement "Every action that falls under no standard moral rule but produces the greatest possible amount of good ought to be performed" contains a descriptive subject-term and a normative predicate-term. Therefore, it should be tested in the same corporatistic way as we test the statement "No act of ingratitude toward one's parents ought to be committed".

I want now to deal with a possible objection. A philosopher might hold that no so-called standard moral rules may be accepted since they all have exceptions and should therefore be jettisoned in favor of a detailed examination of each contemplated action in an effort to see whether the action produces the greatest possible amount of good in the circumstances. My answer to such a philosopher is that standard moral rules are accepted because they help us corporatistically link sensory experience and moral feeling even when they have exceptions. In this respect they resemble descriptive laws of natural science, which are incorporated into the body of science even when sensory experience does not justify their acceptance by the standards of those who think that we must regard each scientific statement as an isolated record of observed fact. Thus a physical formula will sometimes be accepted even though it does not correspond exactly with observations in the laboratory. The actual plotted points on its graph may not yield a straight line when a straight line is called for by the formula, and yet the physicist will often justifiably draw a straight line that misses most or all of the plotted points.[3] Standard rules of morality are acceptable on analogous grounds unless, as it were, the plotted points depart too much from the curve that strictly represents the moral formula. The kind of smoothing and rounding out that goes on in purely descriptive science has its counterpart in normative ethical science. The descriptive scientist tries to link sensory experiences in a simple way whereas the normative ethical scientist will try to link sensory experience and moral feeling in a simple way. Even a large number of standard moral rules may provide a simpler way of organizing the moralist's data than would be pro-

3. Nelson Goodman, *Ways of Worldmaking*, esp. Chapters 1, 6, and 7.

vided by one rule that would require us to calculate the goodness of the causal consequences of each action in order to discover whether we have a duty to perform it.

The fact that such a rule may provide the simplest way of organizing moral data is what justifies our using it in cases where we *lack* standard moral rules. When we use it in such cases we acknowledge that a moral judge who has certain moral feelings and sensory experiences concerning an action which is not subsumable under a standard moral rule needs *some* principle that will help him connect those feelings and experiences. In this respect his thought-processes resemble those of the purely descriptive scientist who needs some principle that will help connect some of his experiences with others. Therefore, the appeal to goodness of consequences and to the principle connecting goodness of consequences with obligation form an intellectual strategy of the moralist which is to be judged in a manner analogous to that in which we judge analogous intellectual strategies of descriptive scientists. And this means that the principle in question is not to be judged by appealing to the self-evidence, a prioriness, analyticity, or logical necessity of such a principle. It might be said that even in cases where standard moral rules are not available, we should *not* appeal to goodness of consequences and to a rule connecting goodness with obligation, on the ground that goodness of consequences has nothing to do with the obligation or right to do something. In answering this we may point out that the epistemic duty of the moralist or the moral agent is to link moral feeling and experience in accordance with certain epistemic rules governing rational thought, and that, in the absence of standard rules, one may appeal to goodness of consequences in order to carry out that duty.

Some other philosophers who oppose the view I advocate may deny that normative statements of ethics and epistemology can be accepted corporatistically because, they hold, the acceptability of "ought"-statements cannot depend on *existent* experiences and moral emotions. In reply to this I should point out that I do not hold that "ought"-statements are logically implied by statements about such sensory experiences and emotions. Also, I cannot be accused of sanctioning what Hume deplored when he criticized the deduction of "ought"-statements from "is"-statements, nor of doing what Moore deplored when he attacked the naturalistic fallacy. Finally, I should point out that if such philosophers hold that we may not test "ought"-statements by seeing whether they play a part in connecting *existent* experiences and moral emotions, how do they think we *can* test such statements? Even if they say that we can do so by examining universals such as Plato's forms or Moore's non-natural attributes, will they not be telling us that in testing "ought"-statements we examine entities that *exist?* If an act is said to be obligatory because it possesses a non-natural attribute, it presumably stands in a certain relationship to something that exists according to their view. Does it not follow, then, that their view is subject to an even stronger objection than the one they level against mine? After all, *they* appeal to existent *universals* whereas I appeal to existent experiences and feelings, in which case I should think that I have the advantage of appealing to entities whose existence is much less doubtful than those to which they appeal.

17. *Epistemological Rules, the Good, and Naturalism*

Since epistemic rules are normative and since I have emphasized the analogies between normative ethics and normative epistemology, one may ask whether the notion of goodness enters the latter in a manner that corresponds to the manner in which it enters the former. Are there circumstances in which standard epistemic rules do not help us in our effort to decide to accept a lower-level body of statements, circumstances in which we then appeal to a concept of *epistemic* goodness in an effort to determine what we ought to accept or reject? When I speak of standard epistemic rules I have in mind, for example, the rule that an accepted body of statements ought to link certain data, that it ought to do so simply, and that it ought to do so while respecting "older truths". But it should be noted that these rules are not usually employed separately, as moral principles are. Often we support a singular moral conclusion by appealing to one moral principle and a relevant descriptive premise, but when we try to support the acceptance of a singular epistemic conclusion, we find ourselves using all of the epistemic rules that I have mentioned simultaneously. We say that the body of statements organizes the data in a simple way while it tips its hat to the "older truths", and when speaking in this manner we simultaneously appeal to *three* rules—the rule of organization, the rule of simplicity, and the rule of conservatism. Nevertheless, the fact that we can meet the demands of these three epistemic rules in varying degrees may lead to a situation in which we must decide how much obedience we should render to each of them. We can imagine a situation in which a simpler theory seriously disturbs our older ways of thinking; we can imagine one in which a simpler theory does some injustice to our sensory experiences; and we can imagine one in which a respect for the older truths also does some injustice to our sensory experiences. In each of these cases there is a conflict of epistemic principles that must be resolved, but how is it to be resolved? By rendering to each of the principles a degree of obedience, so to speak, which strikes a good or satisfactory balance. And this is not unlike the principle that links obligation and good consequences in situations where no standard moral principle is applicable. Here we seem to be resolving an epistemological conflict which is somewhat analogous to a conflict that can occur in normative ethics; and we resolve it by appealing to a notion that is analogous to the notion of ethical goodness. When there is a conflict among principles that govern our acceptance of bodies of statements, we should choose that body of statements which combines the epistemic virtues in the most satisfactory way. This is the principle that corresponds to one that I mentioned in my earlier comments on the role of goodness in ethics. It asserts a connection between what is epistemically satisfactory and what is epistemically obligatory.

I wish to say something further about testing the ethical principle that links goodness and obligatoriness, and also something about the epistemological principle that links two analogous concepts. As usual, I say that they are not tested in isolation but as conjuncts of mixed conjunctions that tie together certain experiences and certain feelings. But, it may be asked, if the word "good" appears in the tested ethical conjunctions and if the word "satisfactory" appears in the tested epistemological conjunctions, and if each conjunction that is accepted ties together certain data, is some feeling especially relevant when we use "ethically good" and another especially relevant when we use "epistemically good" or "epistemically satisfactory"? I believe so. The feeling in each case is that of approval, to which we appeal—with the usual proviso that it is the feeling of a normal person under

normal conditions—when asserting that the consequences of actions that are not governed by standard rules in ethical examples
are good, and also when asserting that we have struck a satisfactory balance in meeting our various epistemic duties. Thus a
first-level moral conjunction will link a feeling of approval with
a feeling of obligation about morally judgeable actions, and a
second-level epistemic conjunction will link a feeling of epistemic
approval with a feeling of epistemic obligation.

The reader will have noticed that in spite of regarding the
term "good" as descriptive rather than as normative, I speak of a
feeling of approval rather than of an experience of any sort.
For we appeal to feelings in assessing arguments containing the
normative word "good". The fact that "good" when used ethically or epistemically is descriptive insofar as it is not used to
state what ought or ought not to be done does not prevent feelings rather than sensory experiences from being relevant here.
The feeling of approval stands to being good in the same relation
as the appearance of red stands to being red.

Having acknowledged the need to appeal in certain circumstances to the concept of good consequences and having also indicated that conjunctions containing words like "good" and "obligatory" are to be tested by consulting feelings of approval and
feelings of obligation, I may well be asked whether I hold a naturalistic theory. In reply, I say that my view is not naturalistic
according to the terminology of certain twentieth-century philosophers since I do not maintain that predicates such as "obligatory" or "good" are *synonymous with* so-called naturalistic predicates. I hope, however, that it is now evident that my view cannot
be characterized as anti-naturalistic either since I do not assert
that predicates such as "obligatory" or "good" are *not* synonymous with naturalistic predicates. In short, I say "A plague o'
both your houses" to reductive ethical naturalists and to their
critics, the ethical anti-naturalists.

I admit, however, that I *may* be called a naturalist if such an
appellation does not rest on any use of the obscure notion of
synonymy. I may be called a naturalist because I believe that
mixed conjunctions in ethics and in epistemology are tested by
seeing whether they link natural items such as sensory experiences and feelings. To be called a naturalist for this reason is
not to commit what is sometimes called the fallacy of deducing

"ought" from "is". So long as I do not assert that a singular normative conclusion—whether it be moral or epistemic—can be deduced from a purely descriptive statement, I do not encourage
the allegedly fallacious deduction of an "ought"-statement from
an isolated "is"-statement. On the contrary, my corporatism shows
that a singular "ought"-statement is deduced from a conjunction
of premises at least one of which contains the word "ought". And
this is certainly compatible with holding that a mixed conjunction of normative and descriptive premises which logically implies a singular "ought"-statement is *tested* by seeing whether the
conjunction links certain experiences and feelings.

Earlier I indicated that I do not appeal to facts conceived as
extra-linguistic abstract entities which we must examine in order
to determine whether a belief that mirrors them is acceptable or
true. But now I want to say something about a related matter.
Philosophers have sometimes contrasted a view like mine with
one according to which a believer ought to square his beliefs with
the state of things as they "really are", and so I feel obliged to
say something about this alternative way of speaking about the
task of the descriptive scientist or that of the systematic moralist.
Now I certainly do not deny the existence of what are called
"real things". But in saying this I do not prevent myself from
saying that a believer ought to link experiences with other experiences or experiences with moral feelings. Since every system
of scientific or moral belief implicitly or explicitly refers to entities that are thought by the believer to be real, there is no dan-
ger that my view of how these systems should be tested will force
a believer to deny the existence of such entities. Since it is a conjunction that is, in my view, tested by its capacity to link data,
those sentences in the conjunction that assert or imply the existence of "real things" will be parts of what is tested. Therefore,
I can heartily agree with what Einstein says about the descriptive
scientist in the following passage: ". . . he appears as *realist* insofar as he seeks to describe a world independent of the acts of
perception; as *idealist* insofar as he looks upon the concepts and
theories as the free inventions of the human spirit (not logically
derivable from what is empirically given); as *positivist* insofar as
he considers his concepts and theories justified *only to* the extent to
which they furnish a logical representation of relations among sensory experiences. He may even appear as *Platonist* or *Pythagorean*

insofar as he considers the viewpoint of logical simplicity as an indispensable and effective tool of his research".[4]

It is difficult, however, for me to say something completely analogous about the moralist. I may of course say in Einstein's terminology that the moralist appears as idealist, as positivist, and as "Platonist or Pythagorean"—even if I add feelings or emotions to the moralist's data—because I believe that the moralist's concepts and theories are free inventions, that he appeals to experience and feelings, and that he values logical simplicity. But do I think that he appears as "realist" in Einstein's sense? No, because I do not think that he always *describes* a world". The premises of a moral argument are not exclusively descriptive because some of them are normative. But when the moralist does make descriptive statements, corporatism does not require the moralist to appeal to a ready-made "fact" in describing an act. The absence of this requirement is a consequence of my contention that the moralist's description of the act is part of a mixed conjunction that is tested by its capacity to link or organize experience *and* moral feeling. The moralist, therefore, believes in the reality of human acts independent of perception but his description of them does not *mirror* a "fact". That is why a moralist may alter or abandon a descriptive statement after having denied the normative conclusion of a moral argument, and do so without being constrained to stick to a previous "mirroring" of a "fact" about the action that he is judging.

What I have just said will be denied by a philosopher who thinks that the descriptive part of the moralist's argument consists of statements that are not free inventions in Einstein's sense. But I do not see how such a philosopher could consistently maintain this while agreeing with Einstein that descriptive statements made by physicists are free inventions. And quite apart from what Einstein says, I do not see how such a philosopher could successfully maintain that the normative principles of the moralist are not free inventions in Einstein's sense. Even if such a philosopher were to point out that moral principles are tenaciously held and held for long periods of time, so long as he granted that a moralist could abandon them or could revise them, such a philosopher would be granting enough for my purposes.

4. Albert Einstein, "Reply to Criticisms", in *Albert Einstein: Philosopher-Scientist,* ed. P. A. Schilpp (Evanston, Illinois, 1949), p. 684.

All I wish to insist on is that the moralist, like the physicist, can alter his body of beliefs in order to organize or link his data in accordance with certain canons. While his normative beliefs will usually be formulated in standard moral rules concerning acts of lying, of stealing, of honoring one's parents, of keeping promises, and so on, there will be occasions on which the moralist will not be able to decide what he ought to do by appealing to such standard rules. In such cases, he may appeal to the principle that he ought to perform the action that leads to the best consequences.

VI

Corporatism Compared
with Some Other Views

18. *Two Defective Versions of the View That
"Ought"-Statements Cannot Be Deduced from
"Is"-Statements: Moore's and Hume's*

Although I pointed out earlier that corporatism does not license
the deduction of "ought"-statements from "is"-statements alone,
I want to distinguish my view of the relationship between such
statements from the influential views of G. E. Moore and David
Hume.

I begin with some brief remarks about Moore's views on the
subject. The Moore of *Principia Ethica* may be called a reductive
anti-naturalist (or non-naturalist) in his treatment of "ought"-
statements because he tries to reduce "ought"-statements to statements that are not naturalistic by way of definition. Moore defines
"duty" in a way that turns every "ought"-statement into a non-
naturalistic statement. "Our duty", he says, "can only be defined
as that action, which will cause more good to exist in the Universe than any possible alternative",[1] and he illustrates his view
by saying that the statement "It is our duty not to murder" means
the same as "Murdering will under no circumstances cause so
much good to exist as the avoidance of murder". Moore must
therefore hold that every such statement about duty is synonymous with one that cannot be tested by the methods of natural
science because he holds that statements which are synonymous

1. *Principia Ethica,* p. 148.

with statements about duty contain the non-natural predicate "good". Of course, I avoid this view because it employs the obscure notion of synonymy as well as the obscure notion of a non-natural predicate, and also because it envisions an epistemological gulf between the testing of normative beliefs and the testing of descriptive beliefs.

Hume is even more famous than Moore for driving a wedge between "ought"-statements and "is"-statements, but Hume's view, as I shall try to show, is more complicated than Moore's. Indeed, I think it is probably inconsistent. Because Hume is a reductive naturalist who holds that "ought"-statements mean the same as "is"-statements, one would not expect him to prohibit the deduction of the former from the latter. Yet he does prohibit such a deduction and, in a manner that I shall analyze, creates a gulf between the testing of normative beliefs and the testing of descriptive beliefs. That is why I think his view suffers from inconsistency, which is reason enough to avoid it. Like Moore, Hume is not an epistemologist of moral belief whose views can be accepted by a corporatist. One adopts a consistent view that is obscure whereas the other adopts a view that is inconsistent.

I have said that by contrast to Moore, Hume is a reductive *naturalist* in his theory of duty or of normative statements. Being a reductionist, as Moore is, Hume tries to tell us what we *mean* by saying that certain acts are moral crimes, namely, that they have causal consequences of a certain kind. But although both Moore and Hume think that we must refer to the causal consequences of actions when saying that they are moral crimes or that we are obliged not to perform them, a crucial difference between Hume and Moore emerges when they specify what these consequences are. Hume is a reductive *naturalist* since he holds that an action is morally criminal because it has the *natural* attribute of producing the sentiment of disapprobation, but Moore is a reductive *anti-naturalist* since he holds that an action is morally criminal because it has the non-natural attribute of not producing as much good as the avoidance of it would. As I have indicated, I do not intend to say much more about Moore's views but I am about to say much more about Hume's, not only because I want to show why I think it is self-contradictory but also to show that his appeal to sentiment or feeling is different from that of corporatism as defended in the preceding pages.

In order to show that Hume is a reductive naturalist, I shall begin with an illustration. Hume is given to calling ingratitude a moral crime, especially ingratitude committed against one's parents; and when Hume says that such ingratitude is a moral crime, he might just as well say "No act of ingratitude towards one's parents ought to be committed".[2] How does Hume think that this statement is established? Since he does not distinguish sharply between calling ingratitude a crime and calling it a vice, we may appeal to Hume's definition of "vice". At one point Hume "defines virtue to be *whatever mental action or quality gives to a spectator the pleasing sentiment of approbation; and vice the contrary*".[3] In accordance with this definition, when people say that ingratitude is a vice or a crime, Hume thinks they *mean* that acts of ingratitude produce the sentiment of disapprobation in a spectator. And it is because Hume defines virtue and vice in this way that I call him a reductive naturalist. He believes that each normative moral statement is synonymous with a psychological statement and he says that whether an act does cause disapprobation in a spectator is a plain matter of fact. In a second definition Hume presents the meaning of the word "vicious" in an equally naturalistic fashion. He says: ". . . when you pronounce any action or character to be vicious, you mean nothing, but that from the constitution of your nature you have a feeling or sentiment of blame from the contemplation of it".[4] And in a third definition Hume defines "virtue" similarly: "It is the nature and, indeed, the definition of virtue, that it is *a quality of the mind*

2. David Hume, *A Treatise of Human Nature*, ed. L. A. Selby-Bigge, p. 466; also David Hume, *An Enquiry Concerning the Principles of Morals*, Appendix I, in *Enquiries Concerning the Human Understanding and Concerning the Principles of Morals*, ed. L. A. Selby-Bigge (2nd edition, Oxford, 1902), p. 287.

In the cited passage of the *Treatise* Hume writes: "Of all crimes that human creatures are capable of committing, the most horrid and unnatural is ingratitude, especially when it is committed against parents, and appears in the more flagrant instances of wounds and death". The following terminological remark is also of interest: "A blemish, a fault, a vice, a crime; these expressions seem to denote different degrees of censure and disapprobation; which are, however, all of them, at the bottom, pretty nearly all the same kind of species", *Enquiry Concerning the Principles of Morals*, Appendix IV, p. 322. However, in at least one place Hume calls ingratitude a vice, *ibid.*, p. 253, n. 4.

3. *Enquiry Concerning the Principles of Morals*, Appendix I, *Enquiries*, p. 289.

4. *Treatise*, p. 469.

agreeable to or approved of by every one who considers or con-templates it".[5] In presenting these definitions Hume plainly implies that when you call an act vicious, criminal, or virtuous, you make a statement which may be established empirically.

The naturalism exhibited in all of these definitions should not be obscured by the fact that the second of them seems more "subjectivistic" than the others. With this in mind, let us compare Hume's first definition of virtue as "whatever mental action or quality gives to *a spectator* [my italics] the pleasing sentiment of approbation" and his first definition of vice as "the contrary" with his second definition, namely, "when *you* pronounce any action or character to be vicious, *you* mean nothing, but from the constitution of *your* nature *you* [all italics mine] have a feeling or sentiment of blame from the contemplation of it". The late C. L. Stevenson once noted that the second definition, which appears in the *Treatise,* seems more subjectivistic than the first definition, offered in the *Enquiry Concerning the Principles of Morals.*[6] The supposed difference would seem to hinge on the repeated use of the words "you" and "yours" in the second definition, by contrast to the use of the phrase "a spectator" in the first. In the first Hume may seem to give a definition of virtue which is "objective" because, according to it, a virtuous action is one that would give *any* spectator (or, as Stevenson suggests with other statements by Hume in mind, *any spectator who is fully informed about the action*) the pleasing sentiment of approbation. And therefore Hume might be thought to hold in the *Enquiry* that all persons mean the same objective thing by "virtuous". On the other hand, the definition in the *Treatise* which contains all of the "you"s might be said to be "subjective" because it seems to imply that Hume is telling at least one of his readers—call him "Smith"—that when Smith says "This action is virtuous", Smith's words mean the same as "This action gives me (Smith) the pleasing sentiment of approbation" whereas when Hume says "This action is virtuous", Hume's words mean the same as "This action gives me (Hume) the pleasing sentiment of approbation".

In my opinion, however, these two definitions do *not* repre-

5. *Enquiry Concerning the Principles of Morals, Enquiries,* p. 261, n. 1. See C. D. Broad, *Five Types of Ethical Theory* (London, 1930), pp. 84-85.
6. *Ethics and Language,* pp. 273-76.

sent two opposed tendencies in Hume's moral philosophy, one "objectivistic" and the other "subjectivistic". In the second one, the one in the *Treatise*, Hume is not saying to his reader: "When *you* pronounce an action or character to be vicious, *you* mean nothing, but that from the constitution of *your* nature *you* have a feeling or sentiment of blame from the contemplation of it; whereas when *I* pronounce an action or character to be vicious, *I* mean nothing, but that from the constitution of *my* nature *I* have a feeling or sentiment of blame from the contemplation of it". What Hume had in mind in the *Treatise* might have been expressed just as well by using "we" where he uses "you" and "our" where he uses "your"; and if he had expressed himself in that way, no one would suspect him of "subjectivism".

To support this view, I want to quote a few sentences that follow the allegedly subjectivistic definition in the *Treatise:* "Vice and virtue, therefore, may be compar'd to sounds, colours, heat and cold, which, according to modern philosophy, are not qualities in objects, but perceptions in the mind: And this discovery in morals, like that other in physics, is to be regarded as a considerable advancement of the speculative sciences; tho', like that too, it has little or no influence on practice".[7] When Hume compares the quality of being a vicious action with the quality of being a white or cold object—whiteness and coldness being what Locke called "secondary qualities"—Hume is not arguing that when Smith says "That action is vicious" Smith means something different from what Hume means when Hume says "That action is vicious". The so-called discovery of secondary qualities did not lead Hume to hold that adjectives expressing these qualities are ambiguous—that Hume meant one thing by "cold" whereas Smith meant another. Therefore, I should reject the view that Hume adopted a subjectivistic definition in the *Treatise*.

However, because Hume's various definitions of "virtue" and "vice" are naturalistic, a serious problem is created by Hume's declaration that "ought"-propositions are not established by reason. He was, as I have said, involved in an inconsistency which does not appear in Moore's theory of "ought"-propositions. Since Hume is a reductive naturalist who defines the terms "virtuous", "vicious", and "criminal" as he does, he *must* regard all "ought"-

7. *Treatise,* p. 469.

statements as synonymous with "is"-statements. In that case, he cannot consistently maintain the impossibility of deducing an "ought"-statement from an "is"-statement and he cannot consistently deny that reason may determine the truth of an "ought"-statement. Soon after Hume remarks that "ought"-propositions cannot be deduced from "is"-propositions, he says that a recognition of this point would "let us see, that the distinction of vice and virtue is not founded merely on the relations of objects, nor is perceiv'd by reason".[8] In other words, Hume does not think that we can establish "ought"-statements by the use of reason. But since Hume is a reductive naturalist in his theory of "ought"-statements, how *can* he maintain that such statements are not established by reason?

In order to answer this question, he asks us to take an action of wilful murder, to examine it in all lights, and to see whether we can find in it "that matter of fact, or real existence, which [we] call *vice*". Continuing, Hume says, "There is no other matter of fact in the case [besides the murder]. The vice entirely escapes you, as long as you consider the object. You never can find it, till you turn your reflection into your own breast, and find a sentiment of disapprobation, which arises in you, towards this action". And then he immediately adds: "Here is a matter of fact; but 'tis the object of feeling, not of reason".[9] So it would appear that a murder, something that we experience with our senses, and a sentiment, something that we feel, are both matters of fact for Hume. In that case the statement "Brutus's act of wilful murder produced a sentiment of disapprobation in Mark Antony" would be a statement that one matter of fact gave rise to, or caused, another matter of fact. But such a statement must be viewed by Hume as one that is established by reason since Hume regards testing by experience as a species of testing by reason.

This interpretation of what Hume the naturalist is obliged to hold concerning normative statements is also borne out by his treatment of another illustration in the first Appendix to his *Enquiry Concerning the Principles of Morals*. There Hume considers a man who is asked why he engages in physical exercise, a man who answers that he does so because he desires to keep his health. When the man is then asked why he desires to keep his

---

8. *Ibid.*, p. 470.
9. *Ibid.*, pp. 468-69. "Murder" seems to be equated with "homicide" here.

health, he answers that it is necessary for the exercise of his calling. And when he is next asked why he is anxious on that account, he replies that he desires to get money. When asked
"Why?" once more, he answers that money is the instrument of
pleasure. If pressed any further, the man, according to Hume,
cannot say any more. Yet Hume himself comments that pleasure
is desirable on its own account and "because of its immediate
accord or agreement with human sentiment and affection".[10] But
is it not an empirical or descriptive statement that pleasure, the
last end which can be offered by the man in the illustration, is
in immediate accord with human sentiment and affection? It is
just as empirical as the statement that exercise is conducive to
health. And because it is, Hume has not escaped the use of reason
or natural science even in the last step of an argument devoted
to showing that the man ought to or at least may use physical
exercise. Hume has reduced the normative statement that a certain act or kind of act ought to, or may, be done to a purely
descriptive statement of the following form: "This act leads to
X; X leads to Y; Y leads to Z; and Z is in immediate accord with
human sentiment and affection". Every conjunct of this conjunction is descriptive and so the conjunction itself is.

I do not deny, of course, that a descriptive or factual statement that exercise leads to health is in *some* way different from
the descriptive or factual statement that pleasure is in "immediate
accord with human sentiment and affection". After all, the former descriptive statement makes no reference to human sentiment and affection. And it is important to stress that Hume holds
that we could never establish that any act or quality is virtuous
without referring to human sentiment and affection. But this contention of Hume is compatible with the view that *we use reason*
in establishing a statement that *refers* to sentiment, for example,
the statement that an action produces the sentiment of approbation in any spectator of a certain kind.

Therefore, it is well to recognize at least two points when
reading Hume on reason, sentiment, and moral knowledge. First
of all, that a statement to the effect that a certain act or kind of
act is virtuous must, in my terminology, be held by Hume to be
normative, synonymous with a descriptive statement *about* the

10. *Enquiries,* p. 293.

relation between an act and the sentiments of a spectator, and therefore establishable by reason. Secondly, that Hume, through a narrow interpretation of the word "descriptive", declines to apply the word "descriptive" to any statement of the form "The act leads to X; X leads to Y; Y leads to pleasure; and pleasure is in immediate accord with human sentiment and affection". This narrow interpretation rests on the idea that it forms no part of the description of an act to say that it leads to something which is in immediate accord with human sentiment and affection. That Hume must have viewed description in such a narrow way is evident from the following effort on his part to support his view of morals by using an analogy. He writes:

> This doctrine will become still more evident, if we compare moral beauty with natural, to which in many particulars it bears so near a resemblance. It is on the proportion, relation, and position of parts, that all natural beauty depends; but it would be absurd thence to infer, that the perception of beauty, like that of truth in geometrical problems, consists wholly in the perception of relations, and was performed entirely by the understanding or intellectual faculties. In all the sciences, our mind from the known relations investigates the unknown. But in all decisions of taste or external beauty, all the relations are beforehand obvious to the eye; and we thence proceed to feel a sentiment of complacency or disgust, according to the nature of the object, and disposition of our organs.
> Euclid has fully explained all the qualities of the circle; but has not in any proposition said a word of its beauty. The reason is evident. The beauty is not a quality of the circle. It lies not in any part of the line, whose parts are equally distant from a common centre. It is only the effect which that figure produces upon the mind, whose peculiar fabric or structure renders it susceptible of such sentiments. In vain would you look for it in the circle, or seek it, either by your senses or by mathematical reasoning, in all the properties of that figure.[11]

On the basis of this passage we may say that Hume regarded description as something that Euclid engages in when he presents a geometric feature of a circle whereas one who speaks of the circle as beautiful does not, according to Hume, assert a descriptive proposition. But I must repeat that to hold, as Hume does,

11. *Ibid.*, pp. 291-92.

that "The circle is beautiful" is equivalent to "The circle produces a certain sentiment upon the mind", is to equate the
former statement with a descriptive causal statement. "The circle
is beautiful" must be a descriptive statement about the circle
according to Hume's analysis of that statement. And if the beauty
of the circle is constituted by the circle's striking the mind in a
certain way, do we not have to use reason to show that the circle
does strike the mind in that way if we want to show that the
circle is beautiful? True, Hume holds that we must support our
assertion by *referring* to the sentiments that any spectator of a
certain kind has when looking at the circle; but our statement
that such a spectator has such sentiments can be tested only by
what Hume calls reason.

Having said this, I should acknowledge that when Hume concludes the first Appendix of his *Enquiry Concerning the Principles of Morals* he says that whereas reason "discovers objects as
they really stand in nature, without addition or diminution",
what he calls "taste" has a "productive" power inasmuch as it
gilds or stains all natural objects "with the colours borrowed from
internal sentiment" and "raises in a manner a new creation".
Consequently, Hume *seems* to hold that reason gives us knowledge of truth about natural objects that are external whereas
taste "makes us feel . . . a new sentiment of blame or approbation"[12] toward that which reason discovers. But this contrast between the roles of reason and taste is compatible with the view
that reason discovers the internal sentiments that taste makes us
feel upon contemplating actions, and with the view that reason
establishes a moral proposition. If reason discovers or establishes
that a natural object—say an act of a certain kind—produces a
favorable sentiment in every person of a certain kind through a
causal chain of whatever length, then reason discovers or establishes that something is virtuous. If it establishes that the same
natural object produces an unfavorable sentiment in every person of that same kind, then it shows that something is vicious.

So long as Hume holds—as he does—that we claim to have
knowledge when we say that something is vicious, he must hold
on the basis of his general epistemology that such a claim either

12. *Ibid.*, p. 294.

concerns relations between ideas or it concerns matters of fact. If it concerns neither, then we must, he holds, commit it to the flames. I am aware that he says that "morals and criticism are not so properly objects of the understanding as of taste and sentiment. Beauty, whether moral or natural, is felt, more properly than perceived". But I am also aware that he defines "virtue" as "whatever mental action or quality gives to a spectator the pleasing sentiment of approbation", and that after so defining "virtue" he immediately adds that we "proceed to examine a plain matter of fact, to wit, what actions have this influence".[13] Yet if the statement that a given action or kind of action has the influence mentioned in the *definiens* of Hume's definition is a statement about a "plain matter of fact", then its *definiendum,* the term "virtue", must be a natural predicate in Moore's sense. I cannot avoid concluding that if Hume, after expressing this view, denies that reason establishes normative propositions, then Hume was inconsistent on the issue. Therefore, I am content to say that I am focusing on one half of his contradictory view, the half in which he held that normative statements are analytically reducible to statements of naturalistic psychology. This half of his contradictory view is incompatible with his view that we cannot deduce "ought"-propositions from "is"-propositions.

In the light of what I have said in this chapter it should be clear why I cannot accept the obscure view of Moore or the contradictory view of Hume on the relationship between "is"-statements and "ought"-statements. Corporatism has the virtue of depending on neither of them. It does not assert that we can deduce "ought"-statements from "is"-statements but neither does it assert that "ought"-statements are tested in a way that is radically different from the way in which we test "is"-statements. It avoids the anti-naturalistic reductionism of Moore and the naturalistic reductionism of Hume. It is especially important to stress that corporatism avoids the naturalistic reductionism of Hume in spite of employing the notion of feeling, emotion, or sentiment. To see this, we must recall how often I have denied that I regard the term "obligatory action" as *synonymous with* any expression that refers to the feelings, emotions, or sentiments of a human

13. *Ibid.,* p. 289.

being. On the other hand, I assert that we defend our normative beliefs by appealing to feelings in a manner that I need not describe once again.

I also want to call attention to another difference between corporatism and Hume's view. Because Hume thinks that we establish or reject descriptive scientific statements in a way that is fundamentally different from the way in which we establish or reject the statement that an act of ingratitude is a moral crime, Hume would not permit us to reject a descriptive scientific statement on the basis of rejecting the statement that an act of ingratitude was in violation of our moral duty. For this reason Hume would not permit the rejection of a descriptive statement that I permitted in my discussion of the mother who killed the fetus. Hume believes that we must test our statements in a one-by-one fashion and therefore does not view moral reasoning in a corporatistic manner. He assigns a very important role to sentiment but that role is very different from the one I assign to it.

19. Corporatism Compared with the Emotive Theory

The late C. L. Stevenson also tried to link science and sentiment in a way that I want to distinguish from the way in which I link them. Stevenson presented an elaborate theory of the manner in which they were linked,[14] and I shall not deal at length with that theory. Yet I do want to say something about his occasional equation of normative sentences with sentences such as "I disapprove of the prisoner's saying yesterday at 4 P.M. 'My regiment went north'; do so as well". According to Stevenson, this sentence has two components, the first of which is descriptive insofar as it reports what the speaker in fact felt and the second of which is not descriptive insofar as it is imperative. Because the second component is an imperative, I do not refer to the sentence of which it is a component as a conjunction. I regard a logical conjunction as a truth-functional compound both of whose components may be true or false; and an imperative may not be said to be true or false. As I understand Stevenson's view when it is illustrated by means of such an example, the first or declarative component expresses what he calls "the cognitive meaning" of a

14. *Ethics and Language, passim.*

normative sentence whereas the second or imperative component may be said to express what he calls its "emotive meaning". Let us therefore call the first component "the descriptive component" and let us call the second "the emotive component".

According to Stevenson, the descriptive component may be supported scientifically whereas the imperative component may not be, just because it *is* an imperative. Imperatives, he holds, cannot be established by deduction or induction. However, Stevenson says, imperatives may be given support that is analogous to the support we give to declarative scientific sentences. If someone asks *why* he should obey the speaker and join him in disapproving of the prisoner's act, the speaker may do things that will *persuade* his interlocutor to join him. For example, he may get the interlocutor to accept a belief that will impel the latter to join the speaker in disapproving of the prisoner's act. Thus if the speaker should say that the prisoner's utterance was a lie, the speaker might succeed in "supporting" the second component— the imperative—which means that he would succeed in getting the interlocutor to obey it and thereby share his, the speaker's, negative attitude toward the act. I should also point out that Stevenson's equation of a normative sentence with a sentence that contains a declarative descriptive component and an imperative emotive component is offered only as a sort of crutch whereby Stevenson can show that the normative sentence has emotive meaning as it stands. Thus, the tone with which the speaker expresses his disapproval and the speaker's gestures may, according to Stevenson, do in a subtle way what the imperative component does in a crude way, namely, express the emotive meaning of his normative statement.

I have said enough to show that Stevenson links science and emotion in a manner different from that in which I link them. I emphasize that Stevenson would illustrate his view by saying that a normative sentence has the same meaning as some other sentence. But, as G. E. Moore once noted,[15] it is not clear how Stevenson uses "the same meaning" here; and this is one reason why I find it difficult to adopt Stevenson's approach. In supposing that he can find the cognitive meaning of ethical sentences he depends on the semantic obscurity upon which other ethical

15. *The Philosophy of G. E. Moore,* ed. P. A. Schilpp, pp. 538-40.

analysts depend. But when he equates a normative sentence with a half-declarative, half-imperative sentence, he introduces a new kind of semantic obscurity. For this equation cannot rest on cognitive synonymy alone but must rest on some kind of cognitive-cum-emotive synonymy that is even more obscure.[16] By contrast, as the reader knows full well by now, I do not try to present cognitive synonyms for ethical sentences and I do not use the notion of emotive meaning. The reader who prefers the view of Stevenson must face the grave difficulties that surround the semantic notions upon which that view rests.

16. The point is that if two sentences are said to have or to lack both the same cognitive meaning and the same emotive meaning, we are saddled with the problem of elucidating two difficult expressions, namely, "same cognitive meaning" and "same emotive meaning". In my view, no one has successfully elucidated these expressions.

VII

Conclusion

20. *Some of the Main Ideas of the Argument*

If I were to try to summarize too much of what I have shown in this study, I would waste the reader's time. However, I do want to mention a few ideas that are central to my argument.

First of all, I have deliberately avoided what is sometimes called the philosophical analysis of normative terms or concepts. I have not tried to say what the word "ought" means, what the word "right" means, or what the word "good" means. Instead, I have concentrated on trying to say how we confirm and disconfirm larger bits of language in which such terms appear. In my opinion, this is a far more important task and one that may be carried out without engaging in the supposedly prior (but quixotic) task of saying what these terms mean. I should add that I do not hold that the method of testing *constitutes* meaning. That is something I have neither held nor tried to defend. I should also add that I have not been operating under the aegis of the Wittgensteinian slogan, "The meaning is the use". I have resolutely tried to philosophize about normative thinking without employing the concept of meaning as conceived by many analytic moral philosophers both dead and living. I hope I have succeeded in dispensing with it. I have also eschewed the notion of emotive meaning as employed by many analytic moral philosophers of the twentieth century.

Second, as the reader well knows, I have applied the doctrine of limited corporatism in my treatment of all kinds of statements. I have applied it to descriptive physical statements which seem to be put on trial in isolation but which ought to be put on trial in conjunction with other statements that are equally descriptive. I have applied it to descriptive statements which seem to be put on trial in isolation but which ought to be tried in conjunction with other statements that are normative. I have applied it to *ethical* normative statements which seem to be put on trial in isolation but which ought to be tried in conjunction with other statements that are descriptive. I have applied it to *epistemic* normative statements which seem to be put on trial in isolation but which ought to be tried in conjunction with other statements that are descriptive. And I have applied the normative doctrine of corporatism to corporatism itself.

Third, when I apply corporatism to a conjunction of statements that are all descriptive statements, I say that the conjunction links sensory experiences with other sensory experiences and I try to explain this notion of linkage or organization both metaphorically and literally. I maintain that a so-called recalcitrant sensory experience may lead us to alter any statement *which has played a part* in this organization or linkage of *certain experiences*. I emphasize "which has played a part" and "certain experiences" to distinguish my view from Quine's view that the *totality* of our beliefs organizes much more of experience. I also believe—although I do not in this study rely especially on this belief—that even a statement of logic or mathematics may be rejected or revised in response to a recalcitrant experience that a scientist may have.

Fourth, when I apply corporatism to a mixed conjunction of statements, some of which are normative and some of which are descriptive, I say that such a conjunction links experiences and feelings, notably the feeling of obligation. Therefore I am led to defend a view that some philosophers may find paradoxical, namely, that we may reject or revise a *descriptive* statement in response to a recalcitrant moral feeling.

Fifth, when I apply corporatism to a mixed conjunction composed of a normative epistemic principle and a descriptive statement about a lower-level body of statements, I refer to a counterpart of the feeling we have about a morally judgeable action, namely, the feeling of obligation we have about a judgeable action such as the acceptance of a lower-level body of statements. I believe that there are such second-level feelings and that they, along with second-level experiences concerning a lower-level body of statements, are linked by a second-level mixed body of statements. Furthermore, recalcitrant second-level feelings may have reverberations in epistemology like those which first-level feelings have in ethics.

I could, of course, add a good deal more by way of summary, but the reader may be pleased that I shall refrain. Instead, I want to say something about certain implications of my summarized views, implications that I have not developed in the earlier parts of this study.

One has to do with the conduct of second-level inquiries into science, for example, with the question: How is the normative epistemology of science related to the description of scientific behavior? I have argued earlier that the normative epistemology of science is a discipline in which we engage in reasoning analogous to normative ethical reasoning. That is to say, it is a discipline in which we employ normative principles and descriptive statements in order to support normative conclusions. But, by the same token, it is also a discipline in which the *testing* of any normative principle should be corporatistic. Therefore, a descriptive statement plays a crucial part in the trial of a normative principle. And this is true because normative principles cannot be tried in isolation unless we repair to intuitions, claims of analyticity, a prioriness, necessity, and other dubious devices of philosophical isolationists. The moral is that the normative epistemology of science and the description of science are yoked in the enterprise of deciding whether to accept a scientific body of statements. In this respect they are like normative ethics and the description of morally judgeable actions. Just as a normative moralist must make descriptive statements about human behavior, so must a normative epistemologist. True, purely descriptive statements about ordinary human behavior and about scientific behavior may also be made and be supported corporatistically in conjunction with *other descriptive statements*. A psychologist of human behavior *need not* describe it with an eye to deciding whether to accept any scientific theory. But the reverse is not true. The testing of normative principles *requires* the assertion of descriptive premises ultimately because—as I have tried to show—normative principles contain descriptive terms. And that is why one cannot be a normative epistemologist of science without being able to say something descriptive about the behavior of scientists.

The corresponding moral for moralists is more evident, but those who have accepted it have not always accepted what corporatism implies, namely, that we have the right to reject a descriptive premise upon rejecting a moral conclusion while reorganizing our experiences and feelings. This shows that our moral feelings may legitimately influence our description of the world and that we must harmonize our experiences and those feelings while justifying any action we contemplate.

LIST OF ILLUSTRATIVE STATEMENTS

(1) Whoever takes the life of a human being does something that ought not to be done.

(2) The mother took the life of a fetus in her womb.

(3) Every living fetus in the womb of a human being is a human being.

(4) The mother took the life of a human being.

(5) The mother did something that ought not to be done.

(6) Every act which is a lie is an act that ought not to be performed.

(6′) Every act which is a lie is an act that a normal person under normal circumstances would feel obligated not to perform.

(7) The prisoner's act of saying yesterday at 4 P.M. "My regiment went north" is a lie.

(7′) The prisoner's act of saying yesterday at 4 P.M. "My regiment went north" is a lie.

(8) The prisoner's act of saying yesterday at 4 P.M. "My regiment went north" is an act that ought not to have been performed.

(8′) The prisoner's act of saying yesterday at 4 P.M. "My regiment went north" is an act that a normal person under normal circumstances would feel obligated not to perform.

(9) Every act that leads to saving the lives of one's countrymen is an act that ought to be performed.

(10) The prisoner's act of saying yesterday at 4 P.M. "My regiment went north" is an act that led to saving the lives of his countrymen.

(11) The prisoner's act of saying yesterday at 4 P.M. "My regiment went north" is an act that ought to have been performed.

(12) Every act which is an act of lying but one that leads to saving the lives of one's countrymen is an act that may be performed, i.e., that we have a right to perform.

(13) The prisoner's act of saying yesterday at 4 P.M. "My regi-

ment went north" is an act of lying but one that leads to saving the lives of his countrymen.

(14) The prisoner's act of saying yesterday at 4 P.M. "My regiment went north" is an act that he had a right to perform.

(15) Every physical event is necessitated.

(16) Every human action is a physical event.

(17) Every human action is necessitated.

(18) If a human action is necessitated, it ought not to be judged morally.

(18') If a human action is necessitated, we ought not to say "It ought to be done", "It ought not to be done", or "It may be done".

(19) No human action ought to be judged morally.

(19') Of no human action ought we to say "It ought to be done", "It ought not to be done", or "It may be done".

Index